Spectroscopy: Modern Concepts

Spectroscopy: Modern Concepts

Edited by **Jason Penn**

New York

Published by NY Research Press,
23 West, 55th Street, Suite 816,
New York, NY 10019, USA
www.nyresearchpress.com

Spectroscopy: Modern Concepts
Edited by Jason Penn

International Standard Book Number: 978-1-63238-426-3 (Hardback)

Printed in the United States of America.

Contents

Preface

Over the recent decade, advancements and applications have progressed exponentially. This has led to the increased interest in this field and projects are being conducted to enhance knowledge. The main objective of this book is to present some of the critical challenges and provide insights into possible solutions. This book will answer the varied questions that arise in the field and also provide an increased scope for furthering studies.

This book provides substantial amount of research on Spectroscopy. Spectroscopy is the field pertaining to absorption and emission of electromagnetic radiation through the interface between matter and energy. It is based on the principle that energy varies on the particular wavelength of electromagnetic radiation. It has proven to be critical as a research instrument in multiple spheres like chemistry, physics, biology, medicine and ecology. The spheres of study are growing rapidly and scientists are delving into upcoming areas in this sphere through the introduction of novel procedures. The motive of this book is to emphasize on the current spectroscopic techniques like Nano Spectroscopy and Optical Spectroscopy. The book also covers organic and physical spectroscopy. The text is a great source of knowledge for education and research functions.

I hope that this book, with its visionary approach, will be a valuable addition and will promote interest among readers. Each of the authors has provided their extraordinary competence in their specific fields by providing different perspectives as they come from diverse nations and regions. I thank them for their contributions.

Editor

Nano Spectroscopy

Photo-Catalytic Degradation of Volatile Organic Compounds (VOCs) over Titanium Dioxide Thin Film

Wenjun Liang, Jian Li and Hong He

Additional information is available at the end of the chapter

1. Introduction

Volatile organic compounds (VOCs) are emitted as gases from certain solids or liquids. VOCs include a variety of chemicals, some of which may have short- and long-term adverse health effects. Concentrations of many VOCs are consistently higher indoors (up to ten times higher) than outdoors.

The control of VOCs in the atmosphere is a major environmental problem. The traditional methods of VOCs removal such as absorption, adsorption, or incineration, which are referred to the new environmental condition have many technical and economical disadvantages. So in recent years, some new technologies called advanced oxidation processes (AOPs), such as biological process, photo-catalysis process or plasma technology, are paid more and more attention.

Advanced oxidation processes (AOPs) are efficient novel methods useful to accelerate the non-selective oxidation and thus the destruction of a wide range of organic substances resistant to conventional technologies. AOPs are based on physicochemical processes that produce in situ powerful transitory species, principally hydroxyl radicals, by using chemical and/or other forms of energy, and have a high efficiency for organic matter oxidation.

Among AOPs, photocatalysis has demonstrated to be very effective to treat pollutants both in gas and in liquid phase. The photo-excitation of semiconductor particles (TiO_2) promotes an electron from the valence band to the conduction band thus leaving an electron deficiency or hole in the valence band; in this way, electron/hole pairs are generated. Both reductive and oxidative processes can occur at/or near the surface of the photo-excited semiconductor particle.

Photocatalytic degradation of VOCs on UV-illuminated titanium dioxide (TiO_2) is proposed as an alternative advanced oxidation process for the purification of water and air. Heterogeneous photo-catalysis using TiO_2 has several attractions: TiO_2 is relatively inexpensive, it dispenses with the use of other coadjutant reagents, it shows efficient destruction of toxic contaminants, it operates at ambient temperature and pressure, and the reaction products are usually CO_2 and H_2O, or HCl, in the case of chlorinated organic compounds. Decomposition path of VOCs with UV/TiO_2 or UV/TiO_2/doped ions is shown in Fig. 1. However, the formations of by-products, such as CO, carbonic acid and coke-like substances, were often observed. These by-product formations are due to low degradation rate of intermediate compounds that are formed by the partial oxidation of VOCs. In order to improve the VOC degradation rate, some authors reported on the enhancement of VOC degradation through the addition of anions (dopant = S, N, P, etc), cations (dopant = Pt, Cu, Mg, etc), polymers or co-doped with several ions on TiO_2, while the difference between doping agents has not been discussed yet.

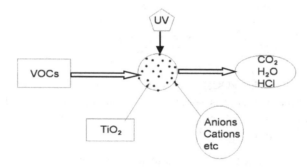

Figure 1. Decomposition path of VOCs with UV/TiO_2.

In this chapter, toluene, p-xylene, acetone and formaldehyde were chosen as the model VOCs because they were regarded as representative indoor VOCs for determining the effectiveness and capacity of gas-phase air filtration equipment for indoor air applications, the photo-catalytic degradation characters of them by TiO_2/UV, TiO_2/doped Ag/UV and TiO_2/doped Ce/UV was tested and compared. The effects of hydrogen peroxide, initial concentration, gas temperature, relative humidity of air stream, oxygen concentration, gas flow rate, UV light wavelength and photo-catalyst amount on decomposition of the pollutants by TiO_2/UV were analyzed simultaneously. Furthermore, the mechanism of titania-assisted photo-catalytic degradation was analyzed, and the end product of the reaction using GC-MS analysis was also performed.

2. Materials and methods

2.1. Chemicals and experimental set-up

Acetone, toluene, p-xylene and formaldehyde used in our experiment was analytical reagent. The TiO_2 photocatalyst was prepared with 100 % anatase using the sol-gel method, and immobilized as a film (thickness 0.2 mm) on glass springs. Ethanol, tetrabutyl orthotitanate, diethanolamine, N,N-dimethylformamide and polyethylene glycol used as raw materials for photocatalyst preparation were of analytical grade and utilized without further purification. $AgNO_3$, $Ce(NO_3)_3 \cdot 6H_2O$ were used as Ag or Ce source for modified TiO_2 samples. Deionized water was used throughout the study.

A schematic diagram of the experimental system for photo-oxidation is shown in Fig. 2. The experiments were performed in a cylindrical photo-catalytic reactor with inner diameter 18.0 mm. A germicidal lamp (wavelength range 200-300 nm) with the maximum light intensity at 254 was installed in the open central region. The desired amount of representative sample, that is acetone, toluene, p-xylene or formaldehyde, was injected into the obturator. Then, the photo-catalytic degradation was performed by transporting the gas across the photo-catalyst continuously when UV lamp was turned on. Glass spring coated by a TiO_2 thin film was filled around the lamp. In whole experiment, humidity was controlled and adjusted with vapour. In some experiments it was replaced with a 15 W black-light lamp with a maximum light intensity output at 365 nm. After a stabilized period of about 3 h, the pollutant concentrations in the outlet gas became the same as in the inlet gas, and the experiment was started by turning on the UV lamp. Relative humidity of the reactor was detected with humidity meter. Oxygen concentration was controlled with oxygen detector.

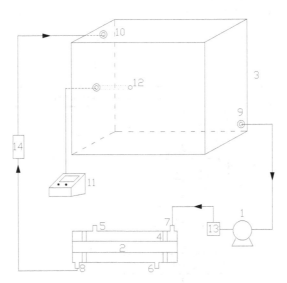

Figure 2. Schematic diagram of experimental set-up. 1- Minitype circulation pump; 2 - Germicidal lamp; 3 - Obturator (airproof tank, 125 L); 4 - Lacunaris clapboard; 5, 6 - Sampling spots; 7-10 – Inlet & Outlet; 11 - Temperature-humidity detector; 12 - Probe; 13 - Gas heated container; 14 - Humidity controller.

2.2. Photo-catalyst preparation

Fig. 3 shows the schematic flow-chart of the experimental procedure.TiO$_2$ precursor sols were prepared by adding tetrabutylorthotitanate (400 mL) into ethanol (960 mL) at room temperature. Then diethanolamine (69.1 mL) was added, and the mixture stirred for 2 hr. Subsequently, ethanol (120 mL), deionized water (25.2 g), 5 wt% AgNO$_3$ or Ce(NO$_3$)$_3$ were added dropwise to the solution. After stirring for 15 min, N,N-dimethylformamide (16.8 mL) was added. This reduced surface tension and made a smooth coating of the thin film on the carrier. The solution was then left to rest for 24 hr. Finally, polyethylene glycol (4.32 g) dissolved in ethanol (120 mL) at 50 °C was added dropwise to the solution. The final solution was left to sit for 12 hr, after which the TiO$_2$ gel had formed. The prepared mixture could remain stable for months at ambient temperature. Thin film TiO$_2$ photocatalyst was formed by dip-coating with a velocity of 5 cm/s. Glass springs were selected as the photocatalyst carrier due to their excellent transparency and long light diffusion distance. Fig. 4 was the sketch of glass spring structure. These were immersed in the TiO$_2$ gel mixture, and then dried at room temperature. This was followed by calcination at 500 °C in a muffle furnace for 2 hr. The glass springs were coated repeatedly (total of five times) using this method to form a thin TiO$_2$ photocatalyst film. The TiO$_2$ film was very stable and durable, and no loss was observed during its application.

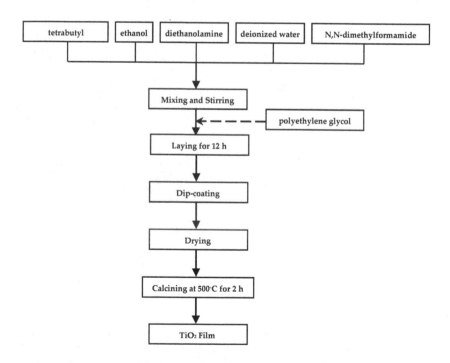

Figure 3. Flowchart of photo-catalyst preparation.

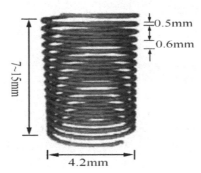

Figure 4. Sketch of glass spring structure.

2.3. Analytical methods

The concentrations of acetone, toluene, and p-xylene were analyzed by a gas chromatograph (Model GC-14C, Shimadzu, Japan) with a flame ionization detector (FID). The oven temperature was held at 60 °C and detector temperature maintained constant at 100 °C. The end products of the reaction were detected by GC-MS. GC-MS analysis was conducted using an HP 6890N GC and HP 5973i MSD. A HP-5 capillary column (30m×0.32mm ID) was used isothermally at 60 °C. The carrier gas (helium) flow-rate was 30 cm/s, and the injector and detector temperatures were 150°C and 280°C, respectively. Intermediate products analysis was done by EI mode and full scan. Formaldehyde concentration in gas stream was determined by acetylacetone spectrophotometric method. HCHO absorbed by deionized water in acetic acideammonium acetate solution would react with acetylacetone to form a steady yellow compound. HCHO concentration in the gas stream was then determined by measuring light absorbance at 413 nm with a spectrophotometer (UV/Vis 722). Temperature and humidity were measured with a temperature-humidity detector (Model LZB-10WB, Beijing Yijie Automatic Equipment Ltd., China).

The characteristics of the immobilized nano-structured TiO_2 thin film were analyzed by field-emission scanning electron microscopy (FE-SEM, Model JSM 6700F, JEOL, Japan) and X-ray diffractometry (XRD, Rigaku, D-max-γA XRD with Cu Kα radiation, λ = 1.54178 Å). The surface area of the TiO_2 film was also analyzed using gas adsorption principles (Detected by Micromeritics, American Quantachrome Co., NOVA 1000). The synthesized samples had a BET surface area of 56.3 m^2/g, compared with Degussa P25 TiO_2 with a surface area of 50.2 m^2/g.

The degradation rates (%) of acetone, toluene, p-xylene and formaldehyde were calculated as follows:

$$\eta = \frac{C_i - C_o}{C_i} \times 100\% \tag{1}$$

where C_i is the inlet concentration, and C_o is the outlet concentration at steady state.

3. Results and discussion

3.1. SEM and XRD Images of the Photocatalyst

FE-SEM analysis of the particle size and shape of the synthesized TiO_2 sample showed it consisted of uniform nano-particles (Fig. 5). However, some cracks were found on the surface. A major contributor to these cracks could be the greater surface tension resulting from the small diameter (0.5 mm) of the glass springs and the high-temperature sintering process. In further experiments, we decreased the temperature from 500 °C to 450 °C. At the lower temperature, there were fewer cracks on the surface but they were not eliminated completely.

 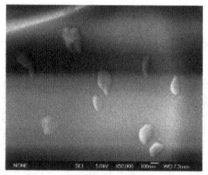

Figure 5. SEM photographs and macroscopic morphology of the TiO_2 thin film coated on a glass spring.

The left and right photographs were taken at 1000× and 50 000× magnification, respectively.

According to Scherrer's equation (Eq. 2) and XRD patterns, the particle size of TiO_2 (D) was calculated to be 35 nm.

$$D = k\lambda / \beta \cos\theta \qquad (2)$$

The crystalline phase of the TiO_2 catalyst was analyzed by XRD (Fig. 6). All the diffraction peaks in the XRD pattern could be assigned to tetragonal anatase TiO_2, with lattice constants of a=0.3785 nm, b=0.3785 nm, and c=0.9514 nm.

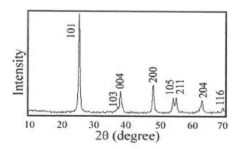

Figure 6. XRD spectrum of anatase crystalline phase of the TiO_2 catalyst

3.2. Effect of doped Ag/TiO₂ or Ce/TiO₂ on decomposition of VOCs

The characters of catalyst are important for the degradation of VOCs. Fig. 7 illustrates the degradation rates of acetone, toluene, and p-xylene (ATP) as functions of irradiation time when pure TiO_2, Ag-TiO_2 and Ce–TiO_2 were used. As controls, blank experiments in the absence of TiO_2 had been studied. The results corresponded to the flow-rate of 1 L/min, initial concentration of 0.1 mol/m³ and relative humidity of 35%. It was found that all the conversions of ATP in the TiO_2/UV, Ag-TiO_2/UV and Ce-TiO_2/UV processes were increased with irradiation time. Table 1 shows the degradation rates for both catalysts after 8-h photo-catalytic reaction. It can be seen from Fig. 7 and Table 1 that the doping of silver or cerium ions could improve the photo-activity of TiO_2 effectively. Furthermore, the degradation character of the photo-catalyst was in the order Ce-TiO_2>Ag-TiO_2>TiO_2. Besides, the results of blank experiments in the absence of TiO_2 showed that the removal efficiency of ATP was very low. For example, the removal efficiency of acetone was merely 6.3% after 8 h and lower than 46.5% for pure TiO_2, which means that TiO_2 plays an important role in photo-catalytic reaction.

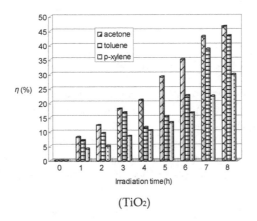

(TiO₂)

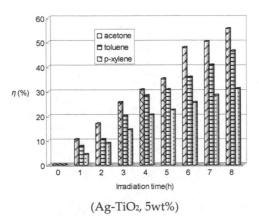

(Ag-TiO₂, 5wt%)

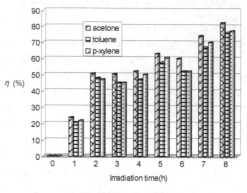

(Ce-TiO$_2$, 5wt%)

Figure 7. Effect of doped Ag/Ce/TiO$_2$ on decomposition of ATP.

Catalyst	TiO$_2$	Ag-TiO$_2$	Ce-TiO$_2$
η (acetone, %)	46.5	55.5	82.0
η(toluene, %)	43.2	46.4	76.2
η (p-xylene, %)	29.8	31.2	77.8

Table 1. ATP degradation rates for different catalysts after 8 hrs.

Fig. 8 illustrated the effect of doped Ag/Ce/TiO$_2$ on decomposition of HCHO. The conditions were as follows: flow-rate of 3 L/min, initial concentration of 0.1 mg/m^3, relative humidity of 35%. It was found that conversions of HCHO in the TiO$_2$/UV, Ag-TiO$_2$/UV and Ce-TiO$_2$/UV processes were increased with irradiation time. It could be seen that the doping of silver or cerium ions could improve the photo-activity of TiO$_2$ effectively. Furthermore, the degradation character of the photo-catalyst was in the order Ce-TiO$_2$ > Ag-TiO$_2$ > TiO$_2$.

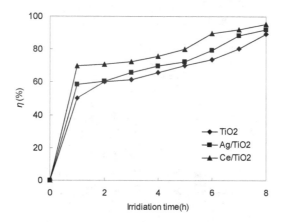

Figure 8. Effect of doped Ag/Ce/TiO$_2$ on decomposition of HCHO.

The reason was as follows: Ag/Ce doping could narrow the band gap. The narrower band gap will facilitate excitation of an electron from the valence band to the conduction band in the doped TiO_2, thus increasing the photo-catalytic activity of the material. At the same time, silver or cerium species could create a charge imbalance, vacancies and unsaturated chemical bonds on the catalyst surface. It will lead to the increase of chemisorbed oxygen on the surface. Surface chemisorbed oxygen has been reported to be the most active oxygen, and plays an important role in oxidation reaction. Herein, silver or cerium modified TiO_2 might have better activity for the oxidation of VOCs. Furthermore, samples after Ag/Ce doping treatment showed a slight change of colour from white to yellowish.

The photo-catalytic activity of Ce-TiO_2 in the oxidative degradation of VOCs being higher than that of Ag-TiO_2 may be explained as follows: Compared to Ag, Ce doping serves as an electron trap in the reaction because of its varied valences and special 4f level. For Ce^{3+}-TiO_2, the Ce 4 f level plays an important role in interfacial charge transfer and elimination of electron-hole recombination. So, Ce doping could enhance the electron-hole separation and the decomposition rate of VOCs could be elevated. Moreover, the valence electrons of TiO_2 catalyst are excited to the conduction band by UV light, and after various other events, electrons on the TiO_2 particle surface are scavenged by the molecular oxygen to produce reactive oxygen radicals. Furthermore, redox reactions between the pollutant molecules and reactive oxygen radicals happened, VOC molecules were turned into harmless inorganic compounds, such as CO_2 and H_2O at the end.

3.3. Effect of Hydrogen Peroxide

Hydrogen peroxide is considered to have two functions in the photo-catalytic degradation. It accepts a photo-generated conduction band electron, thus promoting the charge separation, and it also forms OH•. The addition of H_2O_2 increases the concentration of OH• radicals since it inhibits the electron-hole recombination.

Experiments were conducted to evaluate the effect of H_2O_2 on the toluene/p-xylene photo-degradation. The conditions were as follows: flow rate of 1 L/min, initial concentration of 0.1 mol/m^3, relative humidity of 35%, and photo-catalyst of pure TiO_2. As shown in Fig. 9, the removal efficiency of toluene or p-xylene increased with reaction time.

In the first 3 h, the degradation rate of toluene or p-xylene without H_2O_2 was higher because of the competitive adsorption between toluene or p-xylene molecules and hydrogen peroxide. Then, more reactants and/or radical molecules were produced during the photo-chemistry course, which led to the improvement of toluene or p-xylene decomposition. The final degradation rates of toluene and p-xylene with H_2O_2 were up to 97.1 and 95.4% after 8 h, respectively.

The degradation of acetone was studied with and without hydrogen peroxide (Fig. 10). Overall, the acetone removal efficiency increased with reaction time. Initially, the degradation rate of without H_2O_2 was higher than that of with H_2O_2 because of competitive adsorption between acetone and hydrogen peroxide after hydrogen peroxide addition to the

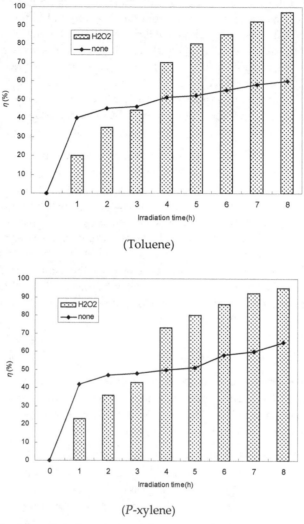

(Toluene)

(P-xylene)

Figure 9. Effect of toluene/p-xylene degradation on hydrogen peroxide.

sample chamber (10 mL per 30 min, 30 % H_2O_2, RH 35 %). As the reactants and/or byproducts accumulated on the catalyst, and there was no new super-oxidation supplied, the catalyst deactivated and the degradation rate increased slowly after 2 hr. Hydroxyl radicals were produced due to the presence of hydrogen peroxide (Eq. 3). This decreased recombination of electron-hole pairs, and consequently the final acetone degradation rate was up to 91.8 % after 8 hr. Consumption of hydroxyl radical likely played an important role in deactivation of the catalyst. An appropriate volume of hydrogen peroxide could enhance the degradation rate, while too much could decrease the degradation rate (Eq. 4).

$$H_2O_2 + e^- \rightarrow \cdot OH + OH^- \tag{3}$$

$$H_2O_2 + \cdot OH \rightarrow H_2O + HO_2 \cdot \tag{4}$$

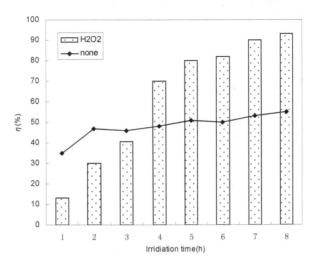

Figure 10. Effect of acetone degradation on hydrogen peroxide.

During deactivation the catalyst color in process without H_2O_2 changed from light white to khaki, while with H_2O_2 was light khaki. This suggests catalyst deactivation in process without H_2O_2 was more extensive than in with H_2O_2. Moreover, the change in catalyst color after the reaction indicates that the reaction occurred in the surface of the catalyst. After sintering at 390 °C for 1 h the catalyst recovered its original color. This again suggests that catalyst deactivation was due to the accumulation of reactants and by-products on the catalyst surface, which impeded degradation reactions. To compare the performance of H_2O_2/UV, we evaluated the potential of acetone degradation by H_2O_2 alone. The concentration of acetone remained almost the same over 8 h. Consequently, we concluded that H_2O_2 alone could not remove the VOCs.

3.4. Effect of initial concentration

In order to discuss the effect of VOCs initial concentration on photo-catalytic degradation rates, we investigated the removal efficiency of ATP and HCHO under different initial concentrations. The ATP concentrations in the experiment ranged between 0.05-0.3 mol/m^3. The conditions were as follows: gas flow-rate of 1 L/min, relative humidity of 35%, Ce-doped TiO_2 as photo-catalyst, and irradiation time of 8 hr. The results showed that the photo-catalytic degradation rates decreased with increasing ATP initial concentration, just illustrated in Fig. 11. Based on the Langmuir-Hinshelwood equation, the degradation rate decreased with increasing initial concentration while the absolute amount of degraded pollutants may increase. At higher initial concentration, the UV light might be absorbed by gaseous pollutants rather than the TiO_2 particles, which led to the reduction of the photo-degradation efficiency. Moreover, at different initial concentrations, acetone was easiest to be destructed, while p-xylene was difficult to be removed among ATP from gas flow.

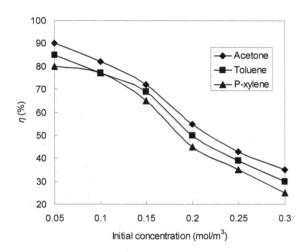

Figure 11. Effect of ATP initial concentration on the photo-catalysis of ATP by TiO_2.

As a main indoor pollutant, the indoor formaldehyde concentration is usually below 0.5 ppmv. It is worth discussing whether the low level of indoor HCHO can be decreased to a value below 0.1 mg/m³ (specified in the indoor air quality standard of China). So in our experiment, the HCHO concentrations in the experiment ranged between 0.1-0.5 mg/m³. The conditions were as follows: relative humidity of 35%, Ce-doped TiO_2 as photo-catalyst, and irradiation time of 120min. The results showed that the photo-catalytic degradation rates decreased with increasing HCHO initial concentration, just illustrated in Fig. 12.

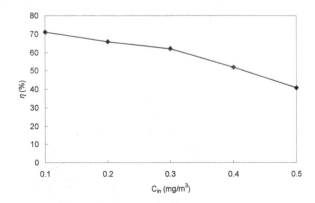

Figure 12. Effect of initial HCHO concentration on HCHO degradation by TiO_2.

In gas-phase photo-catalyst, collision frequency between radicals and HCHO affected the removal efficiency. When formaldehyde molecule reaches to the catalyst surface, the photo oxidation will occur. At higher initial concentration, the UV light might be absorbed by gaseous pollutants rather than the TiO_2 particles, which led to the reduction of the photo-degradation efficiency.

3.5. Effect of UV Light Wavelength

In order to investigate the influence of the UV intensity on the photo-catalytic efficiency, the experiments were performed using two lamp configurations (254 and 365 nm). The effect of UV light wavelength on the efficiency of HCHO degradation is shown in Fig. 13. Just shown in Fig.13, 254 nm UV light provided more effective HCHO photo-degradation than 365 nm UV light.

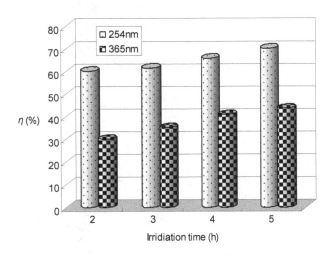

Figure 13. Effect of UV light wavelength on HCHO degradation.

The effect of UV light wavelength on the efficiency of ATP degradation is shown in Fig.14. 254 nm UV light provided more effective ATP photodegradation than 365 nm UV light. Degradation of ATP in the UV/TiO_2 process followed the same trend.

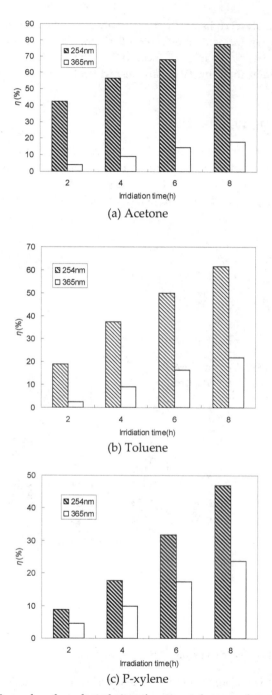

(a) Acetone

(b) Toluene

(c) P-xylene

Figure 14. Effect of UV wavelength on degradation of acetone, toluene, and p-xylene by TiO₂/UV processes.

The different results obtained with 254 and 365 nm UV lamps were mainly due to the stronger UV irradiation from the 254 nm lamp (about 58 W/m² on its surface) than that from the 365 nm lamp (30 W/m² on its surface). This illustrates that the 254 nm UV lamp irradiated photons with higher energy, which led to more efficient degradation with TiO_2/UV.

3.6. Effect of gas flow rate

The effect of gas flow rate on ATP degradation was studied at an initial concentration of 0.1 mol/m³ and relative humidity of 35 %, just as shown in Fig. 15. When the flow rate was increased from 3–9 L/min, degradation of toluene and acetone decreased. With a flow rate >3 L/min the reactants have shorter residence time on the photocatalyst surface and consequently do not bind to the active sites. In general, an increase in gas flow rate results in two antagonistic effects. These are a decrease in residence time within the photocatalytic reactor, and an increase in the mass transfer rate. In our opinion, the decrease in degradation with increasing gas flow rate showed that the residence time of pollutant molecules with TiO_2 is an important factor. However, the degradation rate at 1 L/min was the lowest. This was due to adsorption of active species on the catalyst, which led to a decrease in the reaction between pollutant molecules and active species. For p-xylene, the degradation rate was the highest when the flow rate was 7 L/min. From these results it can be concluded that gas flow rate remarkably influences the degradation rate. While both toluene and p-xylene are aromatic hydrocarbons, toluene is an unsymmetrical molecule and p-xylene is symmetrical. Consequently, the adsorption and degradation of toluene were greater than for p-xylene under the same flow rate. The highest degradation rates for acetone, toluene, and p-xylene were 77.7, 61.9, and 55 %, respectively.

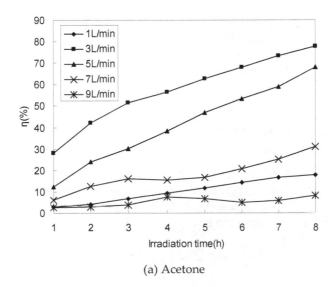

(a) Acetone

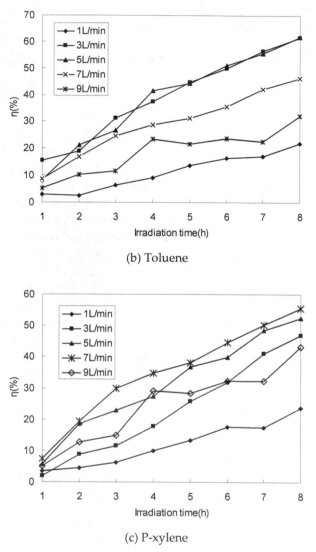

(b) Toluene

(c) P-xylene

Figure 15. Effect of flow rate on the degradation of acetone, toluene, and p-xylene by TiO₂/UV processes.

The Langmuir–Hinshelwood (L–H) rate expression has been widely used to describe the gas–solid phase reaction for heterogeneous photocatalysis. Assuming that mass transfer is not the limiting step, and that the effect of intermediate products is negligible, then the reaction rate in a plug-flow reactor can be expressed as:

$$r = -\frac{dC_{VOC}}{dt} = \frac{kKC}{1 + KC}$$

(5)

where k and K are the L–H reaction rate constant and the L–H adsorption equilibrium constant, respectively; and t is the time taken for ATP molecules to pass through the reactor. After integration of Equation (5) the following linear expression can be obtained:

$$\frac{\ln(C_{in}/C_{out})}{(C_{in}-C_{out})} = \frac{kKT}{(C_{in}-C_{out})} - K \tag{6}$$

where C_{in} and C_{out} are the inlet and outlet concentrations of ATP, respectively; and T is the recurrent time of VOCs in the reactor.

If the L–H model is valid, a plot of $ln(C_{in}/C_{out})/(C_{in}-C_{out})$ versus $1/(C_{in}-C_{out})$ should be linear. This was the case with our data (Fig. 16), and the linearity correlation coefficients of acetone, toluene and p-xylene were 0.9989, 0.9995 and 0.9992, respectively. This result suggests that the reaction occurs on the photocatalyst surface through an L–H mechanism and not in the gas phase.

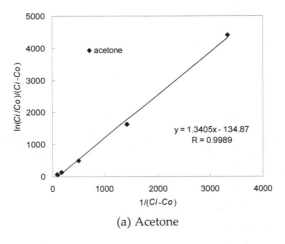

(a) Acetone

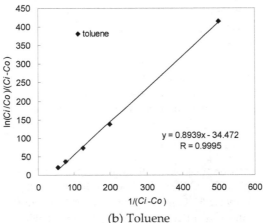

(b) Toluene

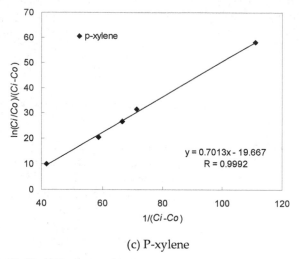

(c) P-xylene

Figure 16. Plot of $\ln(C_{in}/C_{out})/(C_{in}-C_{out})$ and $1/(C_{in}-C_{out})$.

3.7. Degradation of Pure Individual VOCs and Their Mixture

Gaseous-phase photo-degradation for pure individual VOCs (acetone, toluene, and p-xylene) and their mixture was carried out in the continuous flow reactor system. The gas stream passed through the reactor at a flow rate of 5 L/min and contained 0.1 mol/m³ pure acetone, toluene, or p-xylene, or 0.3 mol/m³ of their mixture. The gas residence time was 72 s in the reactor. The experiment was run for 8 hr, and samples were collected at hourly intervals.

Both acetone and p-xylene in the mixed gas degraded at much lower rates than their pure individual gases under the same conditions, just as shown in Fig. 17. However, the opposite trend was observed for toluene. Toluene has an unsymmetrical structure, which leads to instability and promotes adsorption and degradation of pollutant molecules on the catalyst surface according to the L-H mechanism. In addition, the byproducts of acetone and p-xylene produced in the reaction could promote toluene degradation. In contrast, degradation of acetone and p-xylene in the mixed gas was reduced by competitive adsorption and catalysis of toluene. Among the pure gases and the mixture, acetone had the highest degradation efficiency. Furthermore, the efficiency of pure toluene degradation was lower than that of pure p-xylene degradation due to structural stability.

3.8. Effect of gas temperature

The effect of gas temperature on photo-catalytic degradation of gaseous toluene was investigated in the range of 25-50 °C (Fig. 18). The conditions were as follows: gas flow-rate of 1 L/min, relative humidity of 35%, irradiation time of 8 h, photo-catalyst of Ce-doped TiO₂, and initial concentration of 0.1 mol/m³. Degradation efficiency of toluene gradually

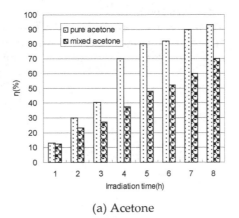

(a) Acetone

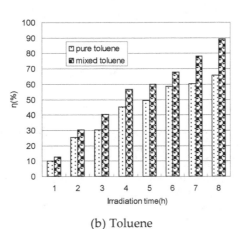

(b) Toluene

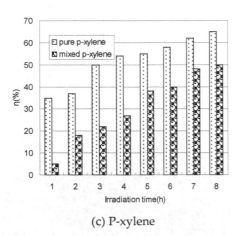

(c) P-xylene

Figure 17. Degradation with H₂O₂ of pure acetone, toluene, p-xylene, and their mixture.

increased when gas temperature was below 45 °C, but decreased at >45 °C. The increase in temperature would lead to the production of free radicals that could effectively collide with toluene molecules. Moreover, higher temperature may increase the oxidation rate of toluene at the interface. However, with increasing temperature, the adsorptive capacities of toluene on catalyst decreased, which led to the reduction of toluene removal efficiency.

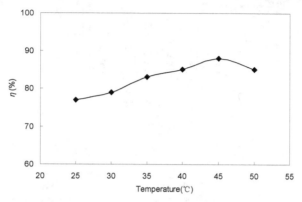

Figure 18. Effect of gas temperature on the photo-catalysis of toluene.

3.9. Effect of photo-catalyst amount

In photo-catalytic degradation of organic compounds, the optimal TiO_2 concentration depends mainly on both the nature of the compounds and the reactor geometry. In this work, the influence of TiO_2 amount on HCHO photo degradation was investigated. A set of gaseous experiments with different amount of TiO_2 from 0 to 100mg was carried out at the RH of 35% and the initial HCHO concentration of 0.1mg/m^3. The degradation rates of HCHO for different amount TiO_2 were presented in Fig. 19. The photo-catalytic degradation efficiency increased with increasing the amount of TiO_2 when TiO_2 amount was lower than 70mg. When the TiO_2 amount was more than 70mg, the photo-catalytic degradation efficiency was decreased. So 70mg of TiO_2 amount was the optional amount in our experiment. And the thickness of 70mg of TiO_2 amount was about 0.2mm.

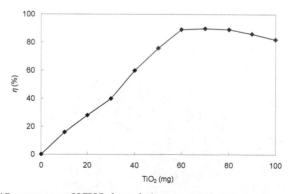

Figure 19. Effect of TiO_2 amount on HCHO degradation

At the same time, in our investigation, the effect of photo-catalyst concentration on the degradation of acetone in the gas flow was also analyzed in order to optimize the amount of TiO_2. Different concentrations (15-105 mg/L) of TiO_2 precursor sols were prepared by using different amounts of tetrabutyl orthotitanate. The conditions of the experiment were as follows: gas flow-rate of 1 L/min, relative humidity of 35%, Ce-doped TiO_2 as photo-catalyst, and irradiation time of 8 h. BET surface area of the synthesized samples was tested (see Table 2). The results showed that BET surface area increased with increasing photo-catalyst amount.

Sample concentration (mg/L)	15	30	45	60	75	90	105
BET (m²/g)	50.2	66.2	68.2	72.2	78.3	88.8	88.9

Table 2. BET surface area for synthesized photo-catalyst.

Fig. 20 showed that the photo-catalytic degradation efficiency increased with increasing the amount of TiO_2. It was suggested that increasing efficiency was due to the increase of the surface area. It could be observed that the degradation efficiency increased with increasing the amount of the catalyst until it reached a plateau at 90-105 mg/L of TiO_2. This indicated that when the amount of TiO_2 was overdosed, the surface area was saturated, and then the intensity of UV was attenuated because of decreased light penetration and increased light scattering.

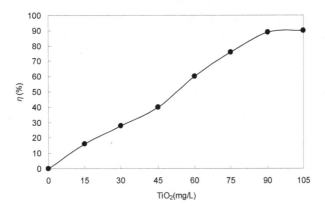

Figure 20. Effect of TiO_2 amount on the photo-catalysis of acetone.

3.10. Effect of relative humidity of air stream

The effect of relative humidity (0-60% RH) of air stream on HCHO decomposition was examined by adding water vapor to a fixed concentration of HCHO. TiO_2 photocatalyst was used in this experiment. Fig. 21 showed the experimental results at different relative humidity. The degradation rate increased with increasing relative humidity up to 35% and then started to decrease, which meant that 35% was the optimal humidity for photo-catalyst process under the experimental conditions. When the reaction time was 120min, the highest removal efficiency of HCHO was 60.2% when RH was 35%.

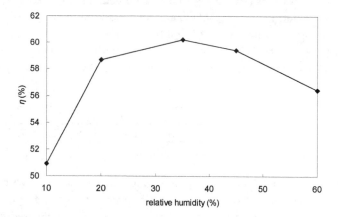

Figure 21. Effect of RH on decomposition of HCHO.

The enhancement of photo-catalytic reaction rate is frequently found in the presence of water vapor because hydroxyl groups or water molecules can behave as hole traps to form surface-adsorbed hydroxyl radicals. In photo-catalyst process, the hydroxyl radicals formed on the illuminated TiO_2 can not only directly attack HCHO molecules, but also suppress the electron-hole recombination. However, higher RH can be attributed to the competition for adsorption between HCHO and hydroxyl radicals, thus decrease the removal efficiency of HCHO.

3.11. Effect of oxygen concentration

The effect of oxygen concentration on HCHO degradation was presented in Fig. 22.

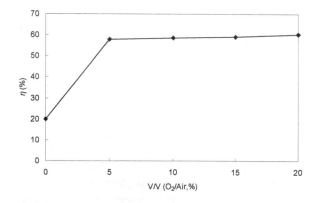

Figure 22. Effect of oxygen concentration on HCHO degradation.

The results corresponded to the initial concentration of $0.1mg/m^3$, relative humidity of 35% and reaction time of 120min. It is obvious that oxidation rates for HCHO increased with increasing O_2 concentration under fixed conditions. As mentioned above, hydroxyl radical is an important factor to the HCHO photo-catalyst. At the same time, oxygen radical is also key factor for HCHO removal, which can react with HCHO on the TiO_2 surface and turn HCHO into CO_2 and H_2O.

3.12. Mechanism of photo-catalytic degradation of VOCs

The heterogeneous photo-catalytic process used in pollutant degradation involved the adsorption of pollutants on the surface sites, and the chemical reactions of converting pollutants into carbon dioxide and water. Activation of TiO_2 is achieved through the absorption of a photon (hv) with ultra-band energy from UV irradiation source. This results in the promotion of an electron (e−) from the valence band to the conduction band, with the generation of highly reactive positive holes (h+) in the valence band. This caused aggressive oxidation of the surface-adsorbed toxic organic pollutants and converts them into CO_2 and water.

In the degradation of toluene or p-xylene, the OH• radicals attack the phenyl ring of toluene or p-xylene, and some products, such as phenol, benzaldehyde or benzoic acid, may be produced during the reaction, and they were converted into CO_2 and H_2O at the end (Fig. 23). We could also observe that acetone was easily destructed to CO_2 and H_2O by photo-catalysts. By-products of toluene or p-xylene were detected by GC-MS, and involved phenol, benzaldehyde, aldehydes, alcohols, etc.

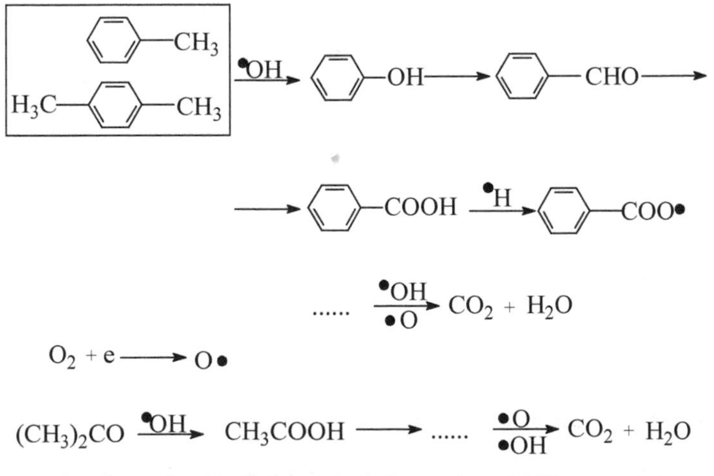

Figure 23. Suggested pathway for the photo-catalytic destruction of ATP.

The reaction rate constant (k) was chosen as the basic kinetic parameter for ATP since it was important in determination of VOCs photo-catalytic activity. The first order kinetic equation:

$$\ln\left(\frac{C_i}{C_o}\right) = k \times t + b \tag{7}$$

was used to fit experimental data in Fig. 24

where C_o is the concentration of ATP remaining in the solution at t, and C_i is the initial concentration at t = 0.

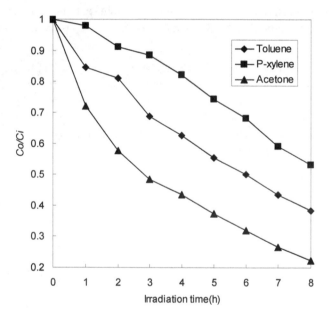

Figure 24. Kinetics of ATP degradation.

The variations in $\ln(C_i/C_o)$ as a function of irradiation time are given in Fig. 25. Reaction rate constant (k), linearity correlation coefficient (R) and intercept (b) data for the photo-catalytic destruction of ATP are exhibited in Table 3. The k of ATP could be ordered as follows: $k_{Acetone} > k_{Toluene} > k_{P-xylene}$, meaning that the decomposition capability of acetone was the best. The reason was probably due to molecular structure and molecular weight.

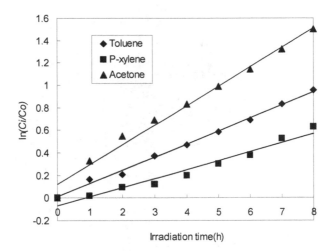

Figure 25. Relation between $\ln(C_i/C_o)$ and irradiation time, and linear fits for ATP.

	k (h^{-1})	R	b
Toluene	0.1165	0.998	0.0102
P-xylene	0.0797	0.980	-0.0668
Acetone	0.1742	0.993	0.12

Table 3. Values of k, R and b for the photo-catalytic destruction of ATP.

During the HCHO decomposition by photo-catalytic processing, formic acid was identified as the intermediate from the photo-degradation of formaldehyde. In our experiment, ion chromatography (IC) was used to determine the byproducts by sampling the gas products into distilled water. The result in this study showed that formic acid was also found. The probably pathway of HCHO destruction was shown in Fig. 26. The related reactions of HCHO destruction were shown with equations (8)-(21).

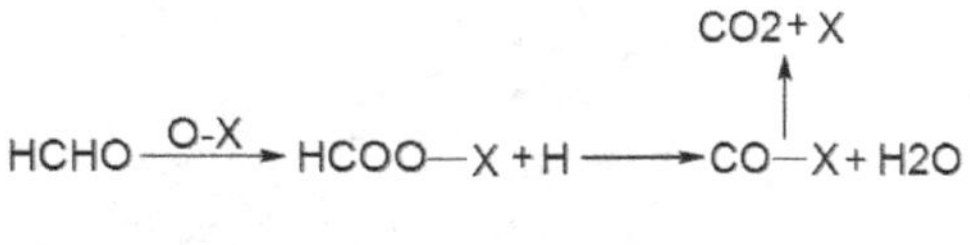

$$CO2+X$$
$$\uparrow$$
$$HCHO \xrightarrow{O\text{-}X} HCOO\!-\!X + H \longrightarrow CO\!-\!X + H2O$$

$$X : TiO_2, \ Ag/TiO_2 \ or \ Ce/TiO_2$$

Figure 26. Suggested pathway for the photo-catalytic destruction of HCHO

$$HCHO + e \longrightarrow H\bullet + CHO^- \tag{8}$$

$$HCHO + O\bullet \longrightarrow OH\bullet + CHO\bullet \tag{9}$$

$$HCHO + OH\bullet \longrightarrow H_2O + CHO\bullet \tag{10}$$

$$HCHO + OH\bullet \longrightarrow HCOOH + H\bullet \tag{11}$$

$$HCHO + H\bullet \longrightarrow H_2 + CHO\bullet \tag{12}$$

$$CHO\bullet + H\bullet \longrightarrow H_2 + CO \tag{13}$$

$$CHO\bullet + O_2 \longrightarrow CO_2 + OH\bullet \tag{14}$$

$$CHO\bullet + OH\bullet \longrightarrow H_2O + CO \tag{15}$$

$$CHO\bullet + HO_2\bullet \longrightarrow OH\bullet + H\bullet + CO_2 \tag{16}$$

$$CHO\bullet + O\bullet \longrightarrow OH\bullet + CO \tag{17}$$

$$CHO\bullet + O\bullet \longrightarrow H\bullet + CO_2 \tag{18}$$

$$HCOOH + OH\bullet \longrightarrow H_2O + H\bullet + CO_2 \tag{19}$$

$$CO + O\bullet \longrightarrow CO_2 \tag{20}$$

$$CO + OH\bullet \longrightarrow CO_2 + H\bullet \tag{21}$$

As mentioned in equation 7, k was the basic kinetic parameter for VOCs photo-catalytic activity. Fig. 27 showed the first order kinetic equation fitting the experimental data.

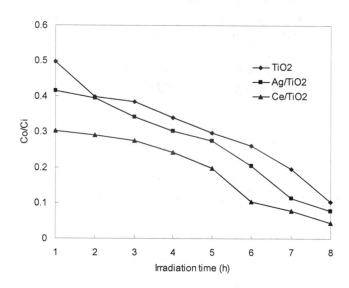

Figure 27. Kinetics of HCHO degradation.

The variations in $\ln(C_i/C_o)$ as a function of reaction time were given in Fig. 28. The reaction rate constant for TiO_2、Ag/TiO_2、Ce/TiO_2 were 0.1871、0.2302、0.2724, respectively, which meant that Ce/TiO_2 had the best photo-catalytic abilities among the catalysts.

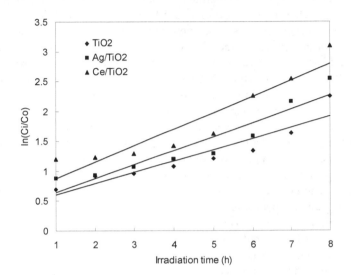

Figure 28. Relationship between $\ln(C_i/C_o)$ and reaction time.

4. Conclusion

In this chapter, nano-structured TiO_2, $Ag-TiO_2$ and $Ce-TiO_2$ thin films coated on glass springs were prepared by sol-gel method at room temperature. Toluene, p-xylene, acetone and formaldehyde were chosen as the model VOCs, the photo-catalytic degradation characters of them by TiO_2/UV, TiO_2/doped Ag/UV and TiO_2/doped Ce/UV was tested and compared. The effects of doped Ag/Ce ions, hydrogen peroxide, initial concentration, gas temperature, relative humidity of air stream, oxygen concentration, gas flow rate, UV light wavelength and photo-catalyst amount on decomposition of the pollutants by TiO_2/UV were analyzed simultaneously. Furthermore, the mechanism of titania-assisted photo-catalytic degradation was analyzed, and the end product of the reaction using GC-MS analysis was also performed.

Results were as follows: (1) Characterization of this film by SEM and XRD showed it consisted of nanoparticles, and the crystalline phase was anatase. (2) Doped Ag or Ce ions could enhance the photo-catalyst ability. The degradation character of the photo-catalyst was in the order $Ce-TiO_2 > Ag-TiO_2 > TiO_2$. (3) Hydrogen peroxide could promote the activation of catalyst, and toluene & p-xylene degradation rate with hydrogen peroxide was higher than that without it. The final degradation rates of toluene and p-xylene using H_2O_2 were up to 97.1 and 95.4% after 8 h, respectively. (4) The photo-catalytic degradation rates decreased with increasing VOCs initial concentration. Acetone was easiest to be destructed, while p-xylene was difficult to remove from gas flow. (5) The degradation efficiency gradually increased with gas temperature and 45 °C had the best removal efficiency. (6) 35% was the optimal humidity for photo-catalyst process under the experimental conditions. (7) Higher concentration of oxygen was better for HCHO removal. (8) The flow rate greatly influenced the degradation rate. For acetone and toluene, the degradation rate was highest with a flow rate of 3 L/min. For p-xylene, the degradation rate was highest when the flow rate was 7 L/min. The highest degradation rates for acetone, toluene and p-xylene were 77.7 %, 61.9 % and 55 %, respectively. (9) Illumination using a 254 nm light source was better than 365 nm. (10) The photo-catalytic degradation efficiency increased with increasing the amount of TiO_2 when TiO_2 amount was lower than 70mg. (11) In the gas mixture, acetone and p-xylene had much lower degradation rates than for their pure counterparts. The opposite trend was observed for toluene. Among acetone, toluene and p-xylene, the removal efficiency of acetone was highest both when pure and as a part of the gas mixture. (12) The photo-catalytic process used in pollutant degradation involved the adsorption of pollutants on the surface sites, and chemical reactions of converting pollutant into CO_2 and H_2O at the end. By-products of toluene or p-xylene were detected by GC-MS analysis, and involved phenol, benzaldehyde, aldehydes, alcohols, etc. The reaction rate constant (k) of ATP was sequenced $k_{Acetone} > k_{Toluene} > k_{P-xylene}$, meaning that the decomposition capability of acetone was the best, probably due to molecular structure and molecular weight. Formic acid was the main byproduct during the decomposition of HCHO. The reaction rate constant (k) of TiO_2 、 Ag/TiO_2 、 Ce/TiO_2 was sequenced $k_{Ce/TiO2} > k_{Ag/TiO2} > k_{TiO2}$, meaning that Ce/TiO_2 had the best photo-catalytic abilities among the catalysts.

Author details

Wenjun Liang, Jian Li and Hong He
College of Environmental and Energy Engineering,
Beijing University of Technology, Beijing, China

Acknowledgement

This work was supported by National high technology research and development program of China (2011AA03A406) and Project of Beijing Municipal Education Commission (KM201110005011).

5. References

Akira F., Tata N.R., Donald A. (2000). Titanium dioxide photocatalysis, Journal of Photochemistry and Photobiology C: Photochemistry Reviews, Vol. 1,pp.1–21.

Alberici, R.M., Jardim, W.E. (1997). Photocatalytic destruction of VOCs in the gas-phase using titanium dioxide. Appl. Catal. B-Environ., Vol.14, pp. 55-68.

Ao C.H., Lee S.C., Yu J.Z., Xu J.H. (2004). Photodegradation of formaldehyde by photocatalyst TiO2: effects on the presences of NO, SO2 and VOCs. Appl. Catal. B., Vol.54, pp.41–50.

Biard, P.F.; Bouzaza, A.; Wolbert, D. (2007). Photocatalytic Degradation of Two Volatile Fatty Acids in an Annular Plug-Flow Reactor; Kinetic Modeling and Contribution of Mass Transfer Rate. Environ. Sci. Technol., Vol.41, pp.2908-2914.

Boulamanti, A. K.; Philippopoulos, C. J. (2008). Photocatalytic degradation of methyl tert-butyl ether in the gas-phase: A kinetic study. J. Hazard Mater., Vol.160, pp. 83-87.

Boulamanti, A.K.; Korologos, C.A.; Philippopoulos, C.J. (2008). The rate of photocatalytic oxidation of aromatic volatile organic compounds in the gas-phase. Atmos. Environ., Vol.42, pp.7844-7850.

Bouzaza A., Vallet C., Laplanche A. (2006) Photocata-lytic degradation of some VOCs in the gas phase using an an-nular flow reactor. Determination of the contribution of mass transfer and chemical reaction steps in the photodegradation process. J. Photochem. Photobiol. A: Chem. Vol.177, pp.212-217.

Carp O., Huisman C.L., Reller A. (2004). Photoinduced reactivity of titanium dioxide. Prog. Solid State Chem., Vol.32, pp. 33-177.

Chatterjee D., Dasgupta S. (2005) Visible light induced photocatalytic degradation of organic pollutants. J. Photo-chem. Photobiol. C. Vol.6, pp.186-205.

Chen F., Pehkonen S.O., Ray M.B. (2002) Kinetics and mechanisms of UV-photodegradation of chlorinated organics in the gas phase. Water Res. Vol.36, pp.4203-4214.

Chen F., Yang Q., Pehkonen S.O., Ray M.B. (2004) Modeling of gas phase photodegradation of chlorinated VOCs. J. Air Waste Manage Assoc. Vol.54, pp.1281-1292.

Chen, X.B.; Mao, S.S. (2007). Titanium Dioxide Nanomaterials: Synthesis, Properties, Modifications, and Applications. Chem. Rev., Vol.107, pp. 2891-2959.

Collins J.J., Ness R., Tyl R.W., Krivanek N., Esmen N.A., Hall T.A. (2001). A Review of Adverse Pregnancy Outcomes and Formaldehyde Exposure in Human and Animal Studies. Regul. Toxicol. Pharm., Vol.34, pp.17-34.

Fujishima A., Zhang X.T. (2006). Titanium dioxide photocatalysis: present situation and future approaches, C. R. Chimie, Vol.9, pp.750–760.

Futamura S., Zhang A., Einaga H., Kabashima H. (2002) Involvement of catalyst materials in nonthermal plasma chemical processing of hazardous air pollutants. Catal. Today. Vol.72, pp.259-265.

González A. S., Martínez S. S. (2008) Study of the sono-photocatalytic degradation of basic blue 9 industrial textile dye over slurry titanium dioxide and influencing factors. Ultrason. Sonochem. Vol.15, pp. 1038-1042.

Hirakawa, T.; Koga, C.;Negishi, N.; Takeuchi, K.; Matsuzawa, S. (2009). An approach to elucidating photocatalytic reaction mechanisms by monitoring dissolved oxygen: Effect of H2O2 on photocatalysis, Appl Catal B: Environ., Vol.87, pp. 46-55.

Holzer F., Roland U., Kopinke F.D. (2002) Combination of non-thermal plasma and heterogeneous catalysis for oxidation of volatile organic compounds Part 1. Accessibility of the intra-particle volume. *Appl. Catal. B: Environ.* Vol.38, pp.163-181.

http://www.epa.gov/iaq/voc.html

Hu, Q.H.; Zhang, C.L.; Wang, Z.R.; Chen, Y.; Mao, K.H.; Zhang, X.Q.; Xiong, Y.L.; Zhu, M.J. (2008). Photodegradation of methyl tert-butyl ether (MTBE) by UV/H2O2 and UV/TiO2. *J. Hazard Mater.*, Vol.154, pp.795-803.

Jacoby W.A., Blake D.M., Noble R.D., Koval C.A. (1995). Kinetics of the Oxidation of Trichloroethylene in Air via Heterogeneous Photocatalysis. *J. Catal.*, Vol.157, pp.87-96.

Jing L.Q., Xu Z.L., Sun X.J., Shang J., Cai W.M. (2001). The surface properties and photocatalytic activities of ZnO ultrafine particles. *Appl. Surf. Sci.*, Vol.180, pp.308-314.

Kecske's T., Rasko'J., Kiss J. (2004). FTIR and mass spectrometric studies on the interaction of formaldehyde with TiO_2 supported Pt and Au catalysts. *Appl. Catal. A*, Vol.273, pp.55-62.

Kim, S.B.; Hong, S.C. (2002). Kinetic study for photocatalytic degradation of volatile organic compounds in air using thin film TiO_2 photocatalyst. *Appl Catal B: Environ.*, Vol. 35, pp.305-315.

Li F.B., Li X.Z., Hou M.F., Cheah K.W., Choy W.C.H. (2005) Enhanced photocatalytic activity of Ce^{3+}-TiO_2 for 2-mercaptobenzothiazole degradation in aqueous suspension for odour control. *Appl. Catal. A: Gen.* Vol.285, pp.181-189.

Li, D.; Haneda, H.; Hishita, S.; Ohashi, N. (2005). Visible-light-driven nitrogen-doped TiO_2 photocatalysts: effect of nitrogen precursors on their photocatalysis for decomposition of gas-phase organic pollutants. *Mater. Sci. Eng. B.*, Vol.117, pp. 67-75.

Li, F.B.; Li, X.Z.; Ao, C.H.; Lee, S.C.; Hou, M.F. (2005).Enhanced photocatalytic degradation of VOCs using Ln^{3+}–TiO_2 catalysts for indoor air purification. *Chemosphere*, Vol.59, pp.787-800.

Lin L., Zheng R.Y., Xie J.L., Zhu Y.X., Xie Y.C. (2007). Synthesis and characterization of phosphor and nitrogen co-doped titania. *Appl. Catal. B.*, Vol.76, pp.196-202.

Liu H.M., Lian Z.W., Ye X.J., Shangguan W.F. (2005). Kinetic analysis of photocatalytic oxidation of gas-phase formaldehyde over titanium dioxide. *Chemosphere*, Vol.60, pp.630-635.

Liu T.X., Li F.B., Li X.Z. (2008). TiO_2 hydrosols with high activity for photocatalytic degradation of formaldehyde in a gaseous phase. *J. Hazard. Mater.*, Vol.152, pp.347-355.

Lu Y.W., Wang D.H., Ma C.F., Yang H.C. (2010). The effect of activated carbon adsorption on the photocatalytic removal of formaldehyde. *Building and Environment*, Vol.45, pp.615–621.

Mohseni, M. (2005). Gas phase trichloroethylene (TCE) photooxidation and byproduct formation: photolysis vs. titania/silica based photocatalysis. *Chemosphere*, Vol.59, pp. 335-342.

Obee, T.N. (1995). TiO_2 Photocatalysis for Indoor Air Applications: Effects of Humidity and Trace Contaminant Levels on the Oxidation Rates of Formaldehyde, Toluene, and 1, 3-Butadiene. *Environ. Sci. Technol.*, Vol.29, pp.1223-1231.

Ogata A., Einaga H., Kabashima H., Futamura S., Kushi-yama S., Kim H. H. (2003) Effective combination of nonthermal plasma and catalysts for decomposition of benzene in air. *Appl. Catal. B: Environ.* Vol.46, pp.87-95.

Ohno T., Tsubota T., Nakamura Y., Sayama K.(2005). Preparation of S, C cation-codoped $SrTiO_3$ and its photocata-lytic activity under visible light. *Appl. Catal. A.*, Vol.288, pp.74-79.

Ohtani B. (2010). Photocatalysis A to Z—What we know and what we do not know in a scientific sense, *Journal of Photochemistry and Photobiology C: Photochemistry Reviews*, Vol.11, pp.157–178.

Olmez T. (2008) Photocatalytic Treatment of phenol with visible light irradiation. *Fresen. Environ. Bull.* Vol.17, pp.1796-1802.

P´erez, M.; Torrades, F.; Garc´ıa-Hortal, J.A.; Dom`enech, X.; Peral, J. (2002). Removal of organic contaminants in paper pulp treatment effluents under Fenton and photo-Fenton conditions. *Appl. Catal. B: Environ.*, Vol.36, pp.63-74.

Parmar, G.R., Rao N.N. (2009). Emerging Control Technologies for Volatile Organic Compounds, *Critical Reviews in Environmental Science and Technology*, Vol.39, pp.41-78.

Sakthivel, S.; Kisch, H. (2003). Photocatalytic and photoelectrochemical properties of nitrogen-doped titanium dioxide. *Chem. Phys. Chem.*, Vol.4, pp. 487-490.

Sano T., Negishi N., Takeuchi K., Matsuzawa S. (2004). Degradation of toluene and acetaldehyde with Pt-loaded TiO_2 catalyst and parabolic trough concentrator. *Solar Energy*, Vol.77, pp.543-552.

Shan G.B., Yan S., Tyagi R.D., Surampalli R.Y., Zhang T.C. (2009). Applications of Nanomaterials in Environmental Science and Engineering: Review, *Practice Peri-odical of Hazardous, Toxic, and Radioactive Waste Management*, Vol.13, pp.110-119.

Shang J., Du Y.G., Xu Z.L. (2002). Photocatalytic oxidation of heptane in the gas-phase over TiO_2.*Chemosphere*, Vol.46, pp.93-99.

Shiraishi F., Yamaguchi S., Ohbuchi Y. (2003). A rapid treatment of formaldehyde in a highly tight room using a photocatalytic reactor combined with a continuous adsorption and desorption apparatus. *Chem. Eng. Sci.*, Vol.58, pp.929-934.

Sleiman, M.; Conchon, P.; Ferronato, C.; Chovelon, J.M. (2009). Photocatalytic oxidation of toluene at indoor air levels (ppbv): Towards a better assessment of conversion, reaction intermediates and mineralization. *Appl Catal B: Environ.*, Vol.86, pp.159-165.

Somekawa S., Kusumoto Y., Ikeda M., Ahmmad B., Horie Y. (2008). Fabrication of N-doped TiO_2 thin films by laser ablation method: Mechanism of N-doping and evaluation of the thin films. *Catal. Commun.*, Vol.9, pp.437-440.

Song L., Qiu R.L., Mo Y.Q., Zhang D.D., Wei H., Xiong Y. (2007). Photodegradation of phenol in a polymer-modified TiO2 semiconductor particulate system under the irradiation of visible light. *Catal. Commun.*, Vol.8, pp.429-433.

Sun R.B., Xi Z.G., Chao F.H., Zhang W., Zhang H.S., Yang D.F. (2007) Decomposition of low-concentration gas-phase toluene using plasma-driven photocatalyst reactor. *Atmos. Environ.* Vol.41, pp.6853-6859.

Tanada S., Kawasaki N., Nakamura T., Araki M., Isomura M. (1999). Removal of Formaldehyde by Activated Carbons Containing Amino Groups. *J. Colloid. Interf. Sci.*, Vol.214, pp.106-108.

Tokumura, M.; Nakajima, R.; Znad, H. T.; Kawase, Y. (2008). Chemical absorption process for degradation of VOC gas using heterogeneous gas–liquid photocatalytic oxidation: Toluene degradation by photo-Fenton reaction. *Chemosphere*, Vol.73, pp. 768-775.

Venkatachalam N., Palanichamy M., Arabindoo B., Mu-rugesan V. (2007). Alkaline earth metal doped nanoporous TiO_2 for enhanced photocatalytic mineralisation of bisphenol-A. *Catal. Commun.*, Vol.8, pp.1088-1093.

Vincenzo A., Marta L., Leonardo P., Javier S. (2006). The combination of heterogeneous photocatalysis with chemical and physical operations: A tool for improving the photoprocess performance, *Journal of Photochemistry and Photobiology C: Photochemistry Reviews*, Vol.7,pp.127–144.

Wang J.H., Ray M.B. (2000) Application of ultraviolet photooxidation to remove organic pollutants in the gas phase. *Sep. Purif. Technol.* Vol.19, pp.11-20.

Wang Y.Y., Zhou G.W., Li T.D., Qiao W.T., Li Y.J. (2009) Catalytic activity of mesoporous $TiO_{2-x}N_x$ photocatalysts for the decomposition of methyl orange under solar simulated light. *Catal. Commun.* Vol.10, pp.412–415.

Wang, X.Q.; Zhang, G.L.; Zhang, F.B.; Wang, Y. (2006). Study integration about Photocatalytic degradation of gaseous benzene in atmosphere by using TiO_2 photocatalyst. *Chemical Research and Application*, Vol.18, pp.344-348.

Wu C.D., Xu H., Li H.M., Chu J.Y., Yan Y.S., Li C.S. (2007) Photocatalytic decolorization of methylene blue via Ag-deposited BiVo(4) under UV-light irradiation. *Fresen. Environ. Bull.* Vol.16, pp. 242-246.

Xu Y.H., Chen H.R., Zeng Z.X., Lei B. (2006). Investigation on mechanism of photocatalytic activity enhancement of nanometer cerium-doped titania. *Appl. Surf. Sci.*, Vol.252, pp.8565–8570.

Yang S.X., Zhu W.P., Jiang Z.P., Chen Z.X., Wang J.B. (2006). The surface properties and the activities in catalytic wet air oxidation over CeO_2–TiO_2 catalysts. *Appl. Surf. Sci.*, Vol.252, pp.8499-8505.

Yang X., Xu L.L., Yu X.D., Guo Y.H. (2008) One-step preparation of silver and indium oxide co-doped TiO_2 photo-catalyst for the degradation of rhodamine B. *Catal. Commun.* Vol.9, pp.1224-1229.

Yi Z., Wei W., Lee S., Gao J.H. (2007). Photocatalytic performance of plasma sprayed Pt-modified TiO_2 coatings under visible light irradiation. *Catal. Commun.*, Vol.8, pp.906-912.

Yu J.C., Yu J.G., Ho W.K., Zhang L.Z. (2001). Preparation of highly photocatalytic active nano-sized TiO_2 particles via ultrasonic irradiation. *Chem. Commun.*, Vol.19, pp.1942-1943.

Yu J.G., Yu J.C., Cheng B., Hark S.K., Iu K. (2003). The effect of F--doping and temperature on the structural and textural evolution of mesoporous TiO_2 powders. *J. Solid State Chem.*, Vol.174, pp.372-380.

Zhang C.B., He H. (2007). A comparative study of TiO_2 supported noble metal catalysts for the oxidation of formaldehyde at room temperature. *Catalysis Today*, Vol.126, pp.345–350.

Zhang C.B., He H., Tanaka K.I. (2005). Perfect catalytic oxidation of formaldehyde over a Pt/TiO_2 catalyst at room temperature. *Catal. Commun.*, Vol.6, pp.211-214.

Zhang C.B., He H., Tanaka K.I. (2006). Catalytic performance and mechanism of a Pt/TiO_2 catalyst for the oxidation of formaldehyde at room temperature. *Appl. Catal. B.*, Vol.65, pp.37–43.

Zhang, P.Y.; Liang, F.Y.; Yu, G.; Chen, Q.; Zhu, W.P. (2003). A comparative study on decomposition of gaseous toluene by O_3/UV, TiO_2/UV and O_3/TiO_2/UV. *J. Photoch. Photobio. A.*, Vol.156, pp.189-194.

Zou, L.D.; Luo, Y.G.; Hooper, M.; Hu, Eric. (2006). Removal of VOCs by photocatalysis process using adsorption enhanced TiO_2-SiO_2 catalyst. *Chem. Eng. Process.*, Vol.45, pp.959-964.

Zuo G.M., Cheng Z.X., Chen H., Li G.W., Miao T. (2006) Study on photocatalytic degradation of several volatile or-ganic compounds. *J. Hazard. Mater.* Vol.128, pp.158-163.

Chapter 2

Spectroscopic Analyses of Nano-Dispersion Strengthened Transient Liquid Phase Bonds

Kavian Cooke

Additional information is available at the end of the chapter

1. Introduction

Transient liquid-phase (TLP) bonding is a material joining process that depends on the formation of a liquid at the faying surfaces by an interlayer that melts at a temperature lower than that of the substrate [1, 2, 3, 4, 5].

The TLP bonding is distinguished from brazing processes by the isothermal solidification of this liquid. This is accomplished since the interlayer is rich in a melting point depressant (MPD). Upon heating through the eutectic temperature, the interlayer will either melt or react with the base metal to form a liquid. During the hold time above the melting temperature, the melting point depressant (solute) into the base metal (solvent). This resulting solid/liquid interfacial motion via epitaxial growth of the substrate is termed "isothermal solidification." A homogeneous bond between the substrates is formed when the two solid/liquid interfaces meet at the joint centerline marking the end of the isothermal solidification stage.

TLP bonding provides an alternative to fusion welding and has been extensively studied for joining particle-reinforced Al-MMCs [1]. The advantage of using this process is that, reinforcing particles are incorporated into the bond region either by using a particle reinforced insert layer or by the melt-back due to a eutectic reaction between the interlayer and the aluminum alloy [6]. It has been shown in the scientific literature that melt-back can be controlled by using thin interlayer materials. It has also been suggested that heating rate, interlayer composition and thickness are most important in reducing melt-back during TLP bonding. These parameters also determine the width of the liquid phase, removal of surface oxide film and particulate redistribution in the bonded region [16, 7,5].

According to Bosco and Zok [8] there exists a critical interlayer thickness at which pore-free bonds are produced. The thickness of the interlayer must exceed that which is consumed through solid-state diffusion; otherwise, no liquid is formed at the bonding temperature. This sets a minimal requirement for the interlayer thickness, which can be determined using equation 1:

$$h_c = h_\eta C_\eta \left(\frac{\rho_\eta}{\rho_i} \right)^2 \tag{1}$$

Where, $2h_\eta$ is the total thickness of the reaction layer formed at the interlayer/MMC interface when the eutectic temperature is reached, C_η is the mass fraction of Ni in the reaction layer. The mass densities of the reaction layer and the Ni interlayer are ρ_η and ρ_{Ni} respectively. On the other hand, Li et al. [5] suggested that a maximum thickness exists that minimizes particle segregation and maximises joint strength. The composition of the interlayer has also been shown by Eagar and McDonald [9] to be equally as important as the interlayer thickness. In the published literature, it has been shown that pure metal interlayers such as nickel [10, 11], silver [12,13] and copper [13, 15] can react with the metal surface by means of eutectic or peritectic reaction to displace surface oxides and regulates the bonding temperature. In joints made using these pure interlayers, the existing problems can be summarized as: parent metal dissolution particulate segregation [13, 14,15], void formation in particle segregated region [12] and low strength micro-bonds at particle-metal interfaces resulting from the poor wettability of molten interlayer on ceramic reinforcement [12,16]. Particle segregation has been shown by Stefanescu et al [17, 18] to be promoted by the slow movement of the solid-liquid interface during isothermal solidification. Li et al. [5] showed that during TLP bonding of Al-MMCs containing particulate reinforcements with diameters less than 30 μm, particle segregation to the interface occurred when a copper foil thickness between 5 and 15 μm was used. Earlier research used alloyed interlayers such as; Zn–Al, Cu–Al and Cu/Ni/Cu systems to decrease bonding temperature in air and to prevent particulate segregation [19]. While composite interlayers such as Al–Si–W mixed powder and Al–Si–Ti–SiC mixed powder were used to improve the densification of a thick reaction layer which formed at the joint. This layer was reinforced with a metallic phase and a ceramic phase [20].

Recent studies into joining Al-MMCs have focused on decreasing bonding temperature by using Sn-based interlayers reinforced with silicon carbide. The results show that joint strength of Al6061 + 25% (Al₂O₃)p improved by approximately 100% when compared to unreinforced Sn-based joints formed by ultrasonic assisted soldering [21]. Yan et al. [22] developed a SiC particle reinforced Zn-based filler which was used to join SiCp/A356 composite. The results indicated that with the use of ultrasonic vibration suitable particle distribution and reduced void formation were achieved. Cooke et al. [23, 24] used an electrodeposited Ni-Al₂O₃ nano-composite coatings to join Al-6061 MMC. The results showed that the use of Ni-Al₂O₃ nano-composite coatings can be used successfully to increase joint strength when compared to TLP bond produced using pure Ni-coating. The results also showed that coating thickness of 5μm can be used to control particle segregation during TLP bonding of Al-6061 MMC

In order to study the kinetics of TLP bonding, techniques such as wavelength dispersive spectroscopy, energy dispersive spectroscopy and x-ray diffraction spectroscopy are normally used, since the compositional changes across the joint region is the mechanism by which the process progresses to completion. In previous studies EDS, WDS and XRD has been used extensive for studying materials due in part to the flexibility that the techniques afford. In most papers, a choice is made between the two processes depending on the information that is required. The difference between these processes are that, the energy-dispersive (ED) type records X-rays of all energies effectively simultaneously and produces an output in the form of a plot of intensity versus X-ray photon energy. The detector consists of one of several types of device producing output pulses proportional in height to the photon energy. Whereas the wavelength-dispersive (WD) type makes use of Bragg reflection by a crystal, and operates in 'serial' mode, the spectrometer being 'tuned' to only one wavelength at a time. Several crystals of different interplanar spacing are needed in order to cover the required wavelength range. Spectral resolution is better than for the ED type, but the latter is faster and more convenient to use. X-ray spectrometers attached to SEMs are usually of the ED type, though sometimes a single multi-crystal WD. This chapter examines the application of spectroscopic analyses such as EDS, WDS and XRD to the evaluation of nanostructure TLP bonds using $Ni-Al_2O_3$ nano-composite thin film as a filler material during TLP bonding of Al-6061MMC. The effects of process parameters on the mechanical and microstructural microstructural changes in the joint region will also be studied.

1.1. Spectroscopic analysis techniques

The characterization of materials using spectroscopic techniques such as energy dispersive spectroscopy or x-ray diffraction spectroscopy techniques is dependent on the generation of a beam of electrons which interacts with the sample to be analyzed. When electrons strike a anode with sufficient energy, X-rays are produced. This process is typically accomplished using a sealed x-ray tube, which consists of a metal target and a tungsten metal filament, which can be heated by passing a current through it resulting in the "boiling off" of electrons from the hot tungsten metal surface. These "hot" electrons are accelerated from the tungsten filament to the metal target by an applied voltage The collision between these energetic electrons and electrons in the target atoms results in electron from target atoms being excited out of their core-level orbitals, placing the atom in a short-lived excited state. The atom returns to its ground state by having electrons from lower binding energy levels make transitions to the empty core levels. The difference in energy between these lower and higher binding energy levels is radiated in the form of X-rays. This process results in the production of characteristic X-rays. X-rays are generated when the primary beam ejects an inner shell electron thus exciting the atom. As an electron from the outer shell drops in to fill the vacancy and de-excite the atom it must give off energy. This energy is specific to each individual element in the periodic table and is also specific to what particular electron dropped in to fill the vacancy. The conversion between energy, frequency, and wavelength is the well-known de Broglie relationship: $E = h\nu = hc/\lambda$, where ν is the frequency, h is Planck's constant (6.62×10^{-34} joule-second), c is the speed of light (2.998×10^{8}m/sec), and λ is the wavelength of the radiation (in m). Based on this relationship, two distinct types of x-ray detector systems are used. These two types of detector systems are called Energy-Dispersive x-ray Spectrometry (EDS) and Wavelength-Dispersive x-ray Spectrometry (WDS).

EDS spectrometer are most frequently attached to electron column instruments such as SEM or (EPMA). As the name implies is a method of x-ray spectroscopy by which x-rays emitted from a sample are sorted out and analyzed based on the difference in their energy level. An EDS system consists of a source of high-radiation; a sample, a solid-state detector (usually from lithium-drifted silicon (Si(Li)); and a signal processing electronics. When the sample atoms are ionized by a high-energy radiation, they emit characteristic x-rays. X-rays that enter the Si(Li) detector are converted into signals (charge pulses) that can be processed by the electronics into an x-ray energy histogram. This x-ray spectrum consists of a series of peaks representative of the type and relative amount of each element in the sample. The number of counts in each peak can be further converted to elemental weight concentration either by comparison standards or standardless calculations. In general, three principal types of data can be generated using an EDS detector: (i) x-ray dot maps or images of the sample using elemental distribution as a contrast mechanism, (ii) line scan data or elemental concentration variation across a given region, and (iii) overall chemical composition, both qualitative and quantitatively.

As the name implies, WDS is a detection system by which x-rays emitted from the sample are sorted out and analyzed based on differing wavelength (λ) in the WDS, or crystal spectrometry. As in EDS or imaging mode, the beam rasters the sample generating x-rays of which a small portion enters the spectrometer. As the fluorescent x-rays strike the analyzing crystal, they will either past through the crystal, be absorbed, be scattered, or be diffracted. Those which satisfy Bragg's Law; $n\lambda = 2d \, Sin\theta$.

(where n = an integer, d = the interplanar spacing of the crystal, θ = the angle of incidence, and λ = x-ray wavelength) will be diffracted and detected by a proportional counter. The signal from this detector is amplified, converted to standard pulse size in the single channel analyzer and counted with a scalar or displayed as rate vs time on rate meter. By varying the positioning crystal one changes the wavelength that will satisfy Bragg's Law. Therefore one can sequentially analyze different elements. By automating crystal movements one can dramatically speed up the analysis time. Typically the WDS analysis is used to gain the same type of information that the EDS is used for, qualitative and quantitative and quantitative information, line scan and dot maps for elemental distribution.

1.2. X-ray diffraction

X-ray diffraction (XRD) is another quantitative spectroscopic technique which reveals information about the crystal structure, chemical composition, and physical properties of materials and thin films. These techniques are based on observing the scattered intensity of an X-ray beam hitting a sample as a function of incident and scattered angle, polarization, and wavelength or energy. Similar to the EDS and WDS techniques discussed above X-ray diffraction is dependent on Bragg equation, which describes the condition for diffraction to occur in terms of the wavelength of the x-radiation (λ), the interplanar ("d") spacings of the crystal, and the angle of incidence of the radiation with respect to the crystal planes (θ). As the spacing between atoms is on the same order as X-ray wavelengths crystals can diffract the radiation when the diffracted beams are in-phase.

When the incident beam satisfies the Bragg condition, a set of planes forms a cone of diffracted radiation at an angle q to the sample. Since the cone of X-rays intersects the flat photographic filmstrip in two arcs equally spaced from the direct X-ray beam, two curved lines will be recorded on the photographic film. The distance of the lines from the center can be used to determine the angle q, which can then be used to determine the interplanar "d" spacing. X-ray powder diffractometers record all reflections using a scintillation detector (in counts per second of X-rays). The pattern of diffracted X-rays is unique for a particular structure type and can be used as a "fingerprint" to identify the structure type. Different minerals have different structure types, thus X-ray diffraction is an ideal tool for identifying different minerals.

2. Electrodeposition of the nano-composite Ni/Al$_2$O$_3$ coating

Composite coatings can be produced by co-deposition of fine inert particles into a metal matrix from an electrolytic or an electroless bath. This technique is getting interesting due to its ability to produce films with excellent mechanical properties such as wear resistance, corrosion resistance, and lubrication.

The preparation of a composite coating is done in two steps. First, an effective dispersion of fine inert particles is produced in the electrolyte. Next, the preparation of the composite coating is made by the manipulation of electrochemical conditions. An effective dispersion of inert particles in the electrolyte promotes the adsorption opportunity of inert particles on the cathode. It causes a higher volume content of inert particles in the composite coating. The mechanical properties of the composite coating are also promoted with the enhancement of the volume content of inert particles in the coating. In electrodeposited composites, the particles to be dispersed in the metal matrix are maintained in suspension in the bath by agitation and they become incorporated in the coating by a process known as electrophoresis.

A number of coating parameters were studied to determine the effect of each on the coating thickness: pH of the electrolyte, current density, electrolyte temperature, agitation frequency and deposition time [33]. Samples were coated for time intervals of 2, 4, 6, 8, 10, 15, 20 and 30 minutes. After electrodeposition, coating thickness was determined using a light microscope. Figure 1(b) shows the relationship between coating thickness and deposition time when; the current density was set to 1.0A/dm^2, pH 3.0, agitation frequency of 300 rpm and an electrolyte temperature of 50°C [25]. It was found that the coating thickness increased with increasing deposition time, in keeping with Faraday's law governing electrode reactions during electrolysis [26].

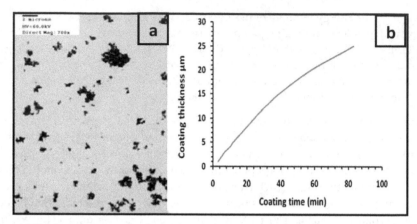

Figure 1. (a) TEM image of as-received nano-sized Al_2O_3 powder (b) Coating thickness as a function of electrodeposition time at 1.0 A/dm^2 and 50°C [33].

In this study, Al -6061/Al_2O_3p samples of dimensions 10 × 10× 5mm were prepared to 800 grit abrasive paper and polished to 1 μm diamond suspension after which they were cleaned in an acetone bath. Acid pickling took place in a solution of 15 wt.% HNO_3 and 2 wt.% HF at 50 °C for 2 minutes and then rinsed in distilled water. The plating solution was prepared by dissolving: 250 g/L $NiSO_4$ $6H_2O$, 45 g/L $NiCl_2.6H_2O$, 35 g/L H_3BO_3 and 1 g/L Saccharin in distilled water. The Ni- Al_2O_3 composite coating was produced by adding 50g/L of ceramic particles to a separately to the nickel bath. The particles were thoroughly mixed into the solution for two hours and kept in suspension in the bath with a magnetic stirrer rotating at 300 rpm. The coating solution was maintained at a temperature of 50°C and pH of 3.0 [33]. The thickness of Ni-Al_2O_3p coatings were controlled by the current density and plating time. The actual amount of Ni-Al_2O_3p electroplated onto a surface was determined by the weight gain after the plating process. The coating thickness was calculated by using the equations 2 and 3:

$$\rho_C = \frac{mass\ of\ coating}{Area\ x\ thickness} = \frac{m}{A\,x\,t} \tag{2}$$

The density of the composite coating (ρ_C) can be calculated by using Equation 3 (rule of mixtures) where x_v is the volume fraction of alumina particles in the Ni-Al_2O_3 coating.

$$\rho_C = \rho_{Al_2O_3} x_v + \rho_{Ni}(1 - x_v) \tag{3}$$

2.1. Spectroscopic analysis of the film structure and composition

Wavelength dispersive spectroscopic (WDS) and X-ray diffraction (XRD) spectroscopic analyses of the coatings deposited were used to evaluate the distribution of the dispersion particles. Figure 2 shows an SEM micrographs of the coating produced by the co-electrodeposition of Ni + 18vol% (nano-Al_2O_3)p. Examination of the coatings using a light microscope revealed the absence of surface defects and interfacial void, however Al_2O_3 particle clusters were present in the coating. This was attributed to particle clustering in the powder prior to the coating process as indicated by the TEM image of the as-received Al_2O_3 powder shown in Figure 1(a). The volume fraction of Al_2O_3 present in the coating was studied by digital x-ray mapping and these results are shown in Figure 3. From Figure 3b and 3c, the areas which contains a high concentration of Al_2O_3 are easily identify and corresponds to Al and O which would confirm the compound to be Al_2O_3 since the base coating is Ni.

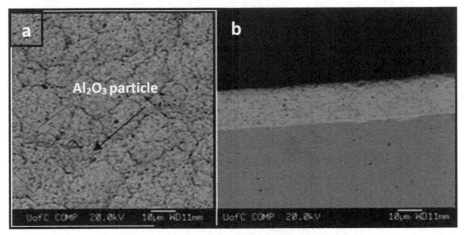

Figure 2. (a) Surface of the Ni-Al_2O_3 coating and (b) Cross-section of the Ni-Al_2O_3 coating produced by electrodeposition [33].

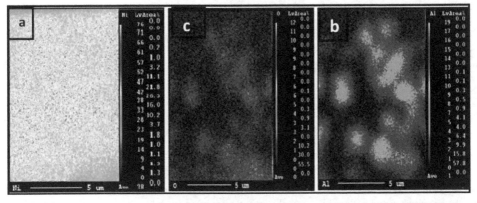

Figure 3. X-ray digital composition maps taken from the Ni-Al_2O_3 coating surface for; (a) Ni, (b) O and (c) Al [33].

3. Fundamentals of transient liquid phase bonding

TLP bonding requires that the base metal surfaces are brought into intimate contact with a thin interlayer placed between the bonding surfaces. The interlayer can be added in the form of a thin foil, powder or coating [27, 28] which is tailored to melt by eutectic or peritectic reaction with the base metal. The liquid filler metal wets the base metal surface and is then drawn into the joint by capillary action until the volume between components to be joined is completely filled.

The driving force of TLP bonding is diffusion. A process which can be described by Fick's first and second laws. The first law describes diffusion under steady-state conditions and is given by equation 4:

$$J = -D\frac{\partial C}{\partial x} \tag{4}$$

Fick's second law describes a non-steady state diffusion in which the concentration gradient changes with time and can be expressed as shown in equation 5:

$$\frac{\partial C}{\partial t} = D\frac{\partial^2 C}{\partial x^2} \tag{5}$$

Equation 6 shows a general solution for Equation-5 using separation of variables is [29]:

$$C(x,t) = \frac{1}{2\sqrt{D\pi t}}\int_{-\infty}^{+\infty} f(\xi)e^{\frac{(\xi-x)}{4Dt}} d\xi \tag{6}$$

Where the error function solution for equation 6 is shown in equation 7:

$$C(x,t) = \frac{C_0}{2}\left[erf\left(\frac{x-x_1}{\sqrt{4Dt}}\right) - erf\left(\frac{x-x_2}{\sqrt{4Dt}}\right)\right] \tag{7}$$

If the following boundary conditions are applied to Equation-7, the concentration as a function of time can be calculated using Equation-8:

$$C(x,0) = \begin{cases} C_o & (x>0) \\ 0 & (0<x) \end{cases} \text{ and } C(x,0) = \begin{cases} C_o & x=+\infty \\ 0 & x=-\infty \end{cases}$$

$$C(x,t) = C_0\left(1 - erf\left(\frac{x}{\sqrt{4Dt}}\right) + erf\left(\frac{kx}{\sqrt{4Dt}}\right)\right) \tag{8}$$

TLP bonding can be conducted by one of two distinct methods. Method-I employs a pure interlayer which forms a liquid through eutectic reaction with the base metal and Method-II employs an interlayer with a liquidus temperature near the bonding temperature [2]. Method-II is most commonly used as it reduces the overall process time by decreasing the volume of solute to be diffused from the interface before the liquid is formed. On the other hand, method-I can be considered to be more effective in TLP bonding as the eutectic reaction is able to displace surface oxides during bonding. TLP bonding process was first divided into five discrete stages by Duvall et al. [30]. These stages were: heating, melting, dissolution of the base-metal, isothermal solidification and homogenization of the excess solute at the bond-line. Zhou later condensed the TLP process to three stages: base-metal dissolution, isothermal solidification and homogenization [7]. In later work by MacDonald and Eager [9] the second and third stage described by Duvall were combined to give four stages; heating, melting and parent metal dissolution, isothermal solidification and homogenization. Zhou [7] reclassified his earlier work to include a heating stage. The second stage was also expanded to be dissolution and widening.

4. Effect of bonding variables on microstructural evolution

This section investigates the effects of bonding variables on the chemical and microstructural homogeneity across the joint region. Factors affecting TLP bonding are; bonding temperature, bonding time, contact pressure and interlayer thickness and composition. Contact pressure can also affect the strength of the bond produced and this has been reported extensive in the scientific literature. However in this work, when bonding pressure increased beyond 0.1 MPa the result is rapid creep failure during the bonding process. Therefore the bonding pressure was held constant at 0.01MPa. The effects of these parameters were evaluated using WDS and XRD to study the change in chemical composition across the joint region as a function of bonding variable.

4.1. Effect of bonding time on joint properties

Wavelength dispersive spectroscopic analyses of joints bonded at 600°C using a 5 μm thick Ni-Al$_2$O$_3$ coating as the interlayer as a function of bonding time as shown in Table 1. For joints bonded for 1 minute a large concentration of Ni remained at the interface after bonding. However, when the bonding time was increased to 30 minutes resulted in the elimination of the interface and an increase in the grain size within the joint. This disrupts the band of segregated particles at the interface (see Figure 4) and chemically homogenized the joint zone.

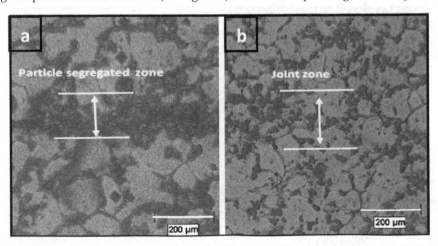

Figure 4. Light micrographs of joint bonded with 5μm thick Ni–Al$_2$O$_3$ coating for: (a) 1 min (b) 30 min [23].

The segregation of particles was accredited to the pushing of micro-Al_2O_3 particles by the solidifying liquid-solid interface. Stefanescu [18] showed that particle pushing can be assumed to be a steady-state condition under which the interface velocity can be assumed to be equal the rate of isothermal solidification. Li et al. [5] suggested that the segregation tendency is dependent on the relationship between the liquid film width produced at the bonding temperature, particle diameter and inter-particle spacing. When the liquid film width is large enough that sufficient particulate material is contained in the melt, particles will be pushed ahead of the solidifying liquid-solid interface resulting in particle segregation at the bond-line. However if the liquid film width is less than some critical value, segregation should not occur.

In the reported studies on transient liquid phase diffusion bonding of Al-MMCs it was shown that the width of the segregated zone at the joint center increased with increasing bonding time. The opposite of this relationship was seen when using the Ni-Al_2O_3 coating. As the bonding time increased, the width of the segregated region decrease. This can be attributed to the heterogeneous nucleation of grains within the joint zone during solidification and this lead to grain refining at the joint. A high resolution SEM micrograph shown in Figure 5 revealed the presence of a nano-sized alumina particle at the center of a grain. EDX spectra of the particle showed Al and O in high concentrations with traces of Mg [23, 24, 33]. Comparing Gibbs free energy of formation at the bonding temperature for MgO (-1195 kJ/mol) and Al_2O_3 (-985 kJ/mol) it is found that Al_2O_3 is unstable in the presence of Mg hence it is like that some of the nano-size Al_2O_3 will decomposed to form $MgAl_2O_4$ compound. WDS analysis across the joint zone as a function of bonding time indicated that the Ni volume at the joint center varied between 3.122 wt% after 1 minute and 0.37 wt% after 30 minutes bonding time (see Table 1).

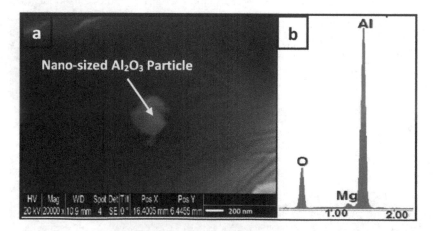

Figure 5. (a) SEM image of nano-particle present in the center of a grain (b) EDS analysis of nano-Al_2O_3 particle [23]

Time (min)	Fe / wt%	Ni / wt%	Si / wt%	Mg / wt%	Cu / wt%	Al / wt%
1	0.399	3.122	0.633	0.379	0.586	93.635
5	0.677	0.849	0.863	1.047	0.71	79.375
10	0.049	0.071	0.211	0.676	0.162	98.598
30	0.028	0.037	0.091	0.458	0.087	99.153

Table 1. WDS analysis of joints made at 600°C as a function of bonding time

4.2. Effect of temperature on joint properties

Wavelength dispersive spectroscopic analyses of the joint as a function of bonding temperature indicated that the Ni concentration at the interface decreased from 4.65 wt% to 0.19 wt% as the bonding temperature is increased from 570 to 620°C (see Table 2). This was attributed to an increase in the diffusivity of Ni from the interface into the base metal as the temperature increased. A review of the scientific literature shows that the diffusivity of Ni increased from D_{570}=4.69 x10^{-13} m²/s to D_{620}= 1.58 x10^{-12} m²/s when the bonding temperature was increased from 570 to 620°C [10, 31].

A study of the joint microstructure for a bond made at 570°C revealed the segregation of Al₂O₃ particles to the bond-line as shown in Figure 6(a). When the bonding temperature was increased to 590°C the width of the particle segregated zone within the joint decreased to approximately 150 μm as shown in Figure 6 (b). Further increase in bonding temperature to 600°C also resulted in a reduction of the width of the segregated zone. A similar result was obtained when the bonding temperature was increased to 620°C (see Figure 6d). This observation was consistent with earlier literature, which suggested that the use of thin interlayers during bonding can help to control the degree of particle segregation taking place within the joint [5]. The micrographs indicate that the width of the particle segregated zone decreases with increasing bonding temperature. This can be attributed to particle pushing by the primary α-phase during solidification as shown in section 4.1 [17, 18].

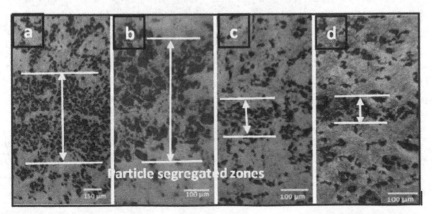

Figure 6. Light micrographs of joint region for bonding temperatures of (a) 570°C and (b) 590°C (c) 600°C and (d) 620°C.

Temperature	Mg/ wt%	Ni/ wt%	Si/ wt%	Fe/ wt%	Al/ wt%
570°C	2.53	4.65	0.72	0.35	91.75
590°C	1.73	1.69	0.52	0.28	95.78
600°C	1.52	0.35	0.41	0.21	97.51
620°C	0.94	0.19	0.21	0.19	98.47

Table 2. Wavelength dispersive spectroscopic analyses of joints made at 600°C for 10 minutes composition as a function of bonding temperature (wt %).

4.3. Effect interlayer thickness on joint properties

The effects of interlayer thickness on microstructural development across the joint region and subsequent effect on joint micro-hardness and shear strength were investigated. According to Bosco and Zok [8], there exists a critical interlayer thickness at which pore-free bonds are produced. This critical interlayer thickness should correspond to maximum joint strength. Therefore the objective in this section is to identify the critical interlayer thickness that maximizes joint strength. Joints made without the use of an interlayer resulted in the formation of a "planar interface" due to the presence of a layer of surface oxide, which prevents metal to metal contact (see Figure 7a). This was corroborated by studies on the solid-state diffusion bonding of Al-MMC [15]. The inability to achieve effective bonding in the solid-state highlights the need for low melting interlayers. When a 1μm thick Ni-Al$_2$O$_3$ coating was used as the interlayer a thin joint zone was achieved (see Figure 7b). However a WDS analysis of this region indicated the presence of a higher concentration of Al$_2$O$_3$ when compared to bonds made using a pure nickel coating of the same thickness. This was attributed to the presence of nano-sized Al$_2$O$_3$ particles in the joint zone and the presence of residual surface oxide.

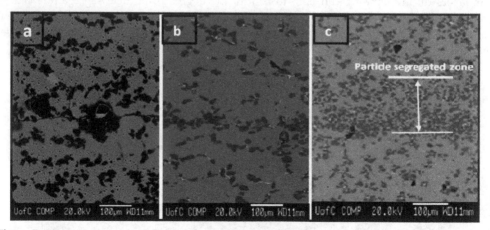

Figure 7. Microstructure of joints bonded at 600°C for 10 min using (a) no-interlayer used (b) 1 μm thick Ni–Al$_2$O$_3$ coating (c) 9μm thick Ni –Al$_2$O$_3$ coating [32].

Figures 8 show that the width of the segregated zone increased with increasing coating thickness. At a coating thickness of 2 μm a more chemically homogeneous joint was created however WDS analysis showed pockets of oxide with the following composition 65.56 wt% and 28.15 wt% were still present at the interface. When the coating thickness was increased to 4 μm, a 95 μm wide segregated zone was at the joint center and the concentration of nickel remaining at the interface after bonding increased from 0.47 wt% with 3μm thick coating to 0.58wt% when a 4μm thick coating (see Table 3). Further increase in coating thickness to 5 μm, resulted in the formation of a 110 μm wide segregated zone while the nickel increased to 0.97 wt%.

Interlayer thickness (μm)	Mg / wt%	Ni / wt%	Si / wt%	Fe / wt%	Al / wt%
1	0.46	0.37	0.53	0.22	98.42
2	0.83	0.49	0.62	0.19	97.87
3	0.98	0.47	0.67	0.21	97.67
4	0.97	0.58	0.72	0.27	97.46
5	0.84	0.97	0.53	0.18	97.48
7	0.86	0.88	0.71	0.38	97.17
9	1.05	1.01	0.69	0.24	97.01
11	1.13	1.45	0.65	0.52	96.25
13	0.99	1.63	0.67	0.97	95.74

Table 3. WDS analysis of joints made at 600°C for 10 minutes as a function of interlayer thickness

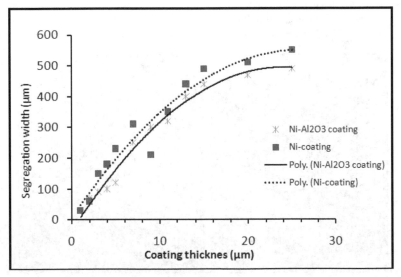

Figure 8. Width of the particle segregated zone formed at within the joint as a function of interlayer thickness for pure Ni coating and Ni–Al₂O₃ coating [32].

The increase in the width of the segregated zone was attributed to increased liquid formation with increasing coating thickness. As the width of the eutectic liquid increases more Al_2O_3 particles are immersed in the liquid phase. These particles are pushed by the solid/liquid interface during isothermal solidification [5].

The width of the particle segregated zone was significantly lower than that achieved when pure Ni-coatings are used as the interlayer. The difference in the width of the segregated zone between joint bonded using pure Ni coating and Ni-Al_2O_3 coating was attributed to the presence of nano-size Al_2O_3 particle in the joint center and a reduction in the concentration of Ni (81.6 wt%) present in the coating, when Ni-Al_2O_3 is used (see Figure 13b).

4.4. Effect of nano-sized particle on joint properties

The effect of nano-sized particles on the microstructural development across the joint region was studied using energy dispersive spectroscopy (EDS). Figure 9(a) shows the microstructure of a joint bonded using a 5μm thick Ni coating dispersed with 50 nm Al_2O_3 particles. From the micrograph a 50 μm wide particle segregated zone was seen within the joint center. Also present at the center of the joint are dark clumps, which EDS analysis suggested are oxide particles (Figure 10 a and b). The presence of the oxide clusters observed, are likely Al_2O_3 particles which agglomerated during the deposition process. When the coating particle size was increased to 500 nm Al_2O_3, a similar result was obtained (see Figure.9b). WDS analyses of the joints as a function of particle size indicated that the Ni concentration of 0.95 wt% and 0.79 wt% for samples bonded using 500 nm and 50 nm respectively. The lower concentration obtained when 50 nm particles are used suggest a faster diffusivity of Ni during the bonding process. This was attributed to greater surface contact between the uncoated Al-6061 sample and the Ni-Al_2O_3 coating surface. Analysis of the roughness using SEM indicated that the surface roughness increased from 0.1 μm for coatings containing 50 nm particles to 0.25 μm for coating containing 500 nm particles.

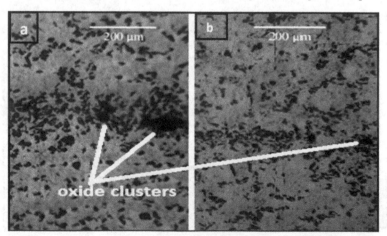

Figure 9. Microstructure of joints bonded at 600°C for 10 min using (a) 5μm thick Ni-(50nm) Al_2O_3 (b) 5μm thick Ni-(500nm) Al_2O_3 [32].

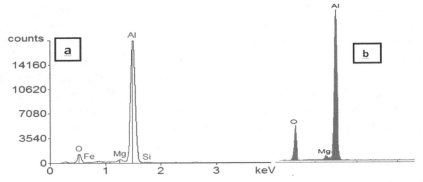

Figure 10. EDS analyses of joints bonded at 600°C for 10 min using (a) 5μm thick Ni-(500 nm) Al₂O₃ (b) 5μm thick Ni-(50 nm) Al₂O₃ [33]

The micrograph shown in Figure 9 a thin segregated zone was formed at the joint center in both cases. The hypothesis is that the difference in the width of the segregated zone obtained is dependent on the differences in particle size, surface roughness of the coating and the distribution of the nano-sized particles in the joint zone during bonding. The presence of nano-Al₂O₃ along the interface was confirmed by TEM analyses which indicated that the nano-particles are arranged along the grain boundary as shown in Figure 11 which would impart a pinning effect as described by Orowan [24].

Figure 11. (a) TEM image of the bonded joint when nano-sized Al₂O₃ particles are used in the interlayer and (b) TEM image of a nano-Al₂O₃ particle located at a grain boundary [33].

5. Effect of bonding variable on the mechanical properties of the joint

5.1. Effect of bonding time the joint shear strength

The shear strength of joints made as a function of bonding time is shown in Figure 12 (a). The graph show that the shear strength increased with increasing bonding time from 68

MPa at 1 minute to 137 MPa at 30 minutes. When a Ni-Al₂O₃ coating was used as the interlayer for a bonding time of 10 minutes, shear strength of 136 MPa was recorded. However when a pure Ni coating was used under that same bonding conditions, shear strength of 117 MPa was achieved [23]. The differences in joint shear strengths obtained were attributed to the presence of a nano-sized dispersion of Al₂O₃ particles within the joint and the precipitation of nickel aluminide phases within the joint region. Shen et al. [34] showed that the increase in yield strength of the nano-particle reinforced aluminum alloy is related to particulate–dislocation interaction by means of the Orowan bowing mechanism. Orowan theory suggests nano-sized particles act as barriers to dislocation motion. This mechanism leads to dislocation pile-up and an increase in the joint shear strength [35].

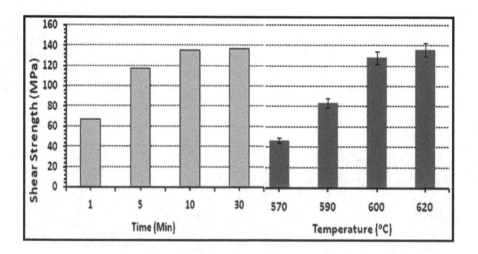

Figure 12. (a) Effect of bonding time on joint strength using 5 μm thick Ni- Al₂O₃ coating at a bonding temperature of 600°C and (b) Effect of bonding temperature on joint shear strengths made using 5 μm thick Ni-Al₂O₃ coatings for 10 minutes [23, 24].

The fractured surfaces of the shear tested joints were analyzed to identify the mechanism of joint failure. Figure 13(a) shows the fractured surface of a bond made for a bonding time of 1 minute. The fractograph shows an undulating surface containing shear plastic deformation with some cleavage facets indicating a mixed failure mode. The fracture appeared to have propagated through the bond-line. The result of the fractographic analyses suggests that the mechanism of failure transitioned from brittle to ductile as the bonding time increases. When the bonding time was increased to 10 minutes (see Figure 18b). The surface was characterized by an undulating appearance of plastic deformation indicative of ductile mode of failure. Additionally, fractured micro-Al₂O₃ particles were observed at the fractured surface, indicating a transgranular fracture through the bond-line.

XRD analyses of the fractured surfaces of bonds made at 1 and 10 minutes are shown in Figure 14. The results indicated the presence of peaks for phases such as $AlFe_6Si$ ($2\theta=38°$), Al_9FeNi and $AlFeSi$. The literature showed that these ternary compounds forms readily in Al-Mg-Si-Fe-Ni systems through various peritectic reactions [17]. In addition, binary crystal phases such as Al_3Ni, Ni_3Si ($2\theta=78°$). Al_3Si and Al_2O_3 compounds were also identified.

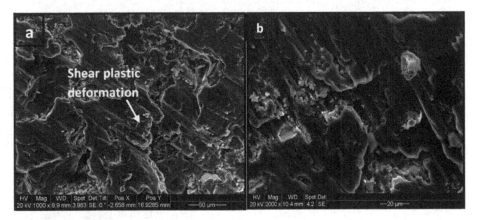

Figure 13. SEM micrograph the fractured surface of a bond made at 600°C with 5 μm thick Ni–Al$_2$O$_3$ for: (a) 1 minute and (b) 10 minutes

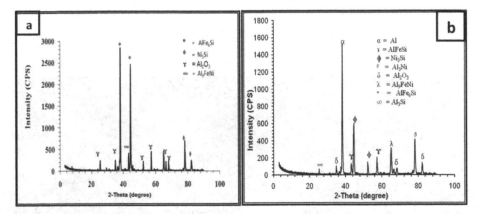

Figure 14. XRD analysis of the fractured surface of a bond made with 5 μm thick Ni–Al$_2$O$_3$ for (a) 1 minute and (b) 10 minutes and

5.2. Effect of temperature on joint shear strength

Joint shear strengths measured as a function of bonding temperature were obtained using a single lap shear test. A comparison of the joint shear strengths of bonds made at 570, 590,

600 and 620°C is shown in Figure 12(b). The test result show that the shear strength increased with increasing bonding temperature from 45 MPa at 570°C to 138 MPa 600°C. This increase in joint strength was attributed to the presence of nano-sized ceramic particles and the precipitation of intermetallic phases within the joint region. The formation of these nickel aluminide phases increased with increasing bonding temperature. Specimens bonded at 620°C gave the highest bond strength of 136 MPa. The effect of the nano-particles and the precipitated intermetallics on the composite was discussed by Zhang and Chan [45] and results in Orowan strengthening as discussed in the previous sections.

In order to compare the effect of bonding temperature on joint failure mechanisms the fractured surfaces were examined using SEM. The results collected suggested that the ductility of the joint increased with increasing bonding temperature. For a bonding temperature of 570°C a mix failure mode was observed with both shear rupture dimples and cleavage planes (Figure 15a). An XRD analysis of the fractured surface indicated a high concentration of Al_2O_3 particles (see Figure 16a). This indicated that at this temperature the matrix-particle (M-P) interface was the weakest point for crack propagation giving the lowest joint strength of (53 MPa). Fracture propagation was observed through the bond-line. When the bonding temperature was increased to between 590°C and 620°C XRD analyses of the fracture indicated that the amount of intermetallic formed within the joint increased to include the binary compounds Al_3Si, Al_3Ni and Ni_3Si which suggested that failure propagated through the bond-line. Additionally, peaks of the ternary phases AlFeSi ($2\theta=37$ and 68°) and Al_9FeSi ($2\theta=78$ and 84°). The presence of these compounds confirms the formation of peritectic reactions during bonding. At 620°C the fractured surface was characterized by shear ruptured dimples indicating a ductile failure mode which occurred in the parent metal adjacent to the bond-line.

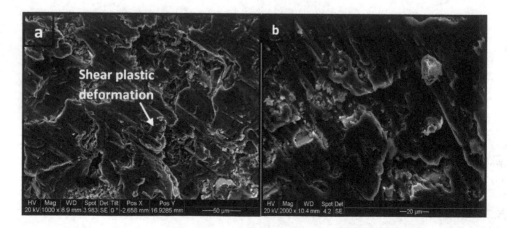

Figure 15. (a) SEM micrograph of the fractured surface of a joint made using a 5 μm thick Ni-Al₂O₃ coatings at 570°C and (b) 620°C.

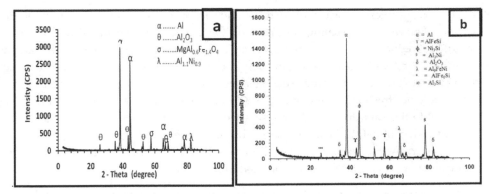

Figure 16. XRD spectrum of the fractured surface of a joint made using a 5 μm thick Ni-Al₂O₃ coatings at (a) 570°C and (b) 620°C.

5.3. Effect of coating thickness on shear strength measurements

Figure 17 (a) shows the variation in joint shear strength values as a function of coating thickness. The graph shows that the shear strength increased with increasing coating thickness from 53 MPa at 1 μm to 144 MPa at 11 μm. This increase in joint strength was attributed to three phenomena: the presence of nano-size Al₂O₃ particles in the joint center, the segregation of micro Al₂O₃ particle to the joint zone and the precipitation of intermetallic phases such as Al₃Si, Ni₃Si, Al₃Ni, and Al₉FeNi within the joint region. As discussed in the previous sections, for composites containing nano-sized particles, strengthening is often explained by the Orowan mechanism [36, 37]. Orowan bypassing theory shows that when smaller particle reinforcements are used the result is a more effective pinning of dislocation motion compared to when micro-particles are used. This mechanism leads to an increase in joint strength and hardness.

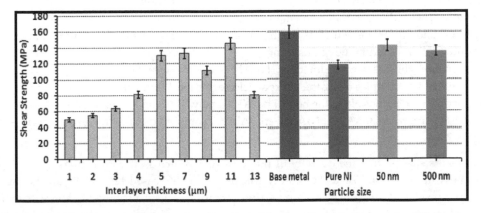

Figure 17. (a) Joint shear strengths as a function of particle size using 5μm thick coatings (b) Shear strength profile plotted as a function of Ni-Al₂O₃ coating thickness.

When the coating thickness was increased beyond 11 μm, a decline in the strength of the joints was observed. At a coating thickness of 13 μm, a joint strength of 80 MPa was recorded. This reduction in joint strength was attributed to the formation of densely packed micro-Al_2O_3 particle-rich regions along the bond interface and also due to an increase in the volume of intermetallics compounds such as $AlFe_3Si$ within the joint. The literature shows that the volume fraction of the micro-Al_2O_3 particles within the joint is inversely proportional to the joint ductility. Therefore, as the width of the particle segregated zone increased the ductility of the joint decreases. This leads to embrittlement of the joint region and causes a reduction in joint strength [38,39, 40, 41]. The findings published in the scientific literature supports the results collected in this study.

Fractured surfaces of the shear tested joints were analyzed to identify the mechanism of joint failure and the composition of the fractured surfaces. Figures 18 shows the micrographs of the typical fractured surfaces obtained for joints that were bonded using coating thickness ranging from 1 to 3μm, respectively. Fractographic analyses revealed that the fractured surface contained cleavage planes, which propagated through the bond-line. In addition, Al_2O_3 particles were visible at the fractured surface. Examination of the fractured surfaces revealed characteristics of a brittle fracture which suggest that insufficient eutectic liquid is formed when using coating thickness between 1 and 3μm. Composition analysis of the fractured surfaces using XRD indicated the presence of peaks for Al_2O_3, Ni_3Si and $Ni_{17}Al_{3.9}Si_{5.1}O_{48}$ compounds.

When an interlayer thickness of 11μm (see Figure 19a) was used, the fractured surface showed evidence of both shear plastic deformation and fractured micro- Al_2O_3 particles indicating a ductile transgranular fracture. Samples bonded at this condition had the highest shear strength. This indicated a critical combination of segregated micro- Al_2O_3 particles and nano-Al_2O_3. Crack propagation occurred in the base metal adjacent to the bond-line. The results suggest that within coating thickness range of 5 to 11μm, sufficient eutectic liquid is produced, which facilitate good particle to matrix bonding resulting increased joint strength. The XRD spectrum shown in Figure 19 (b) indicated the presence of Ni_3Si, $MgAl_2O_4$ and Al_2O_3 compound at the fractured surface. Increasing the coating thickness beyond 11μm resulted in the gradual transition of the fracture mode from ductile to brittle. At a coating thickness of 13μm thick Ni-Al_2O_3 coating (see Figure 20), the surface is characterized by dimples along the interparticle regions indicating a ductile failure through the particle-rich regions along the bond-line. XRD analyses of the fractured surfaces indicate the presence of peaks for Al_2O_3, $NiAl_2O_4$ and $AlFe_3Si$ compound at the fractured surface (see Figure 20b).

The results suggest that the ductility of the joint region increased with increasing coating thickness up to 9 μm. When the coating thickness was increased beyond 9 μm the joint region transitioned from ductile to brittle. These transitions were attributed to an increase in the volume of eutectic liquid that forms with increasing coating thickness leading to interparticle contact.

Figure 18. (a) SEM micrograph (b) cross-section (Mag. X10) and (c) XRD analysis of the fractured surface for a bond made with a 3 μm thick Ni–Al₂O₃ coating for 10 minutes at 600°C.

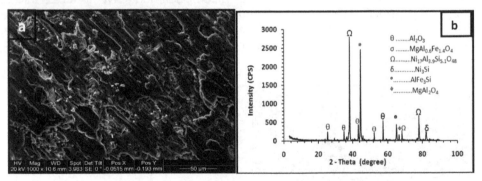

Figure 19. (a) SEM micrograph (b) cross-section (Mag. X10) and (c) XRD analysis of the fractured surface for a bond made with an 11 μm thick Ni–Al₂O₃ coating for 10 minutes at 600°C.

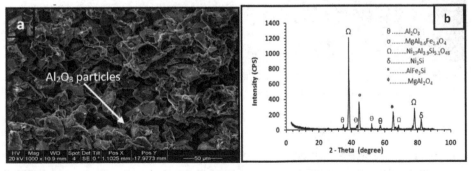

Figure 20. (a) SEM micrograph and (b) XRD analysis of the fractured surface for a bond made with a 13 μm thick Ni–Al₂O₃ coating for 10 minutes at 600°C.

5.4. Effect of interlayer particle size shear strength measurements

Figure 17(b) show the joint shear strength graph as a function of interlayer particle size. The result indicated that joint shear strength increased with decreasing particle size from 138

MPa with 500 nm to142 MPa with 50 nm. This increase in joint shear strength was attributed to better distribution of nano-sized particles within the interlayer when smaller particle sizes are used. In both cases higher shear strengths were obtained than when pure Ni coating is used (117 MPa) [23]. The results indicate that joint strength of up to 90% that of the base metal (BM) is achievable when using a 50 nm diameter nano-sized particle-reinforced interlayer. Tjong [42] showed that the nano-particle size has a strong effect on the yield strength. The author suggested that a particle size of 100 nm is a critical value for improving the yield strength of nano-composites. Below this critical value the yield strength increases significantly with decreasing particle size. Similar results were obtained by Gupta and coworkers [43, 44]. Zhang and Chen [45] showed that the Orowan stress plays a major role in strengthening the nano-composites.

Figure 21 shows the fractured surface for a bond made using a 5μm Ni- 50 nm Al_2O_3 particle. The fractograph showed shear plastic deformation, indicative of ductile fracture with a crack propagating primarily through the bond-line and a section of the base metal adjacent to the bond-line. A similar result was obtained when a dispersed particle size of 50 nm were used in the coating. XRD analyses of the fractured surfaces indicated the presence of peaks for Al_2O_3, $NiAl_2O_4$ and $Al_{11}Ni_9$ compound at the fractured surface.

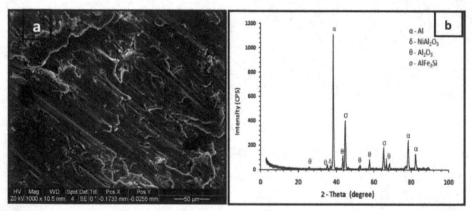

Figure 21. (a) SEM micrograph and (c) XRD analysis of the fractured surface for a bond made with 5 μm thick Ni–(50nm) Al_2O_3 for 10 minutes at 600°C.

6. Mechanism of joint formation

The mechanism of bond formation using composite Ni-Al_2O_3 coating is discussed in this section. The experimental results and the scientific literature show that the joint formation during transient liquid phase diffusion bonding is completed in five distinct stages: interfacial contact and solid-state diffusion, eutectic melting and base metal dissolution, and isothermal solidification. These stages will be discussed thoroughly with reference to the change in composition across the joint region. Mathematical models for predicting the parameter settings when nano-composite coatings are used in joining will also be presented. The sub-section of this topic are as follows:

6.1. Interfacial contact and Solid-state diffusion

The first stage of transient liquid phase bonding involved heating the sample to the bonding temperature. During this stage of bonding, two mechanisms are thought to occur simultaneously, firstly an increase in interfacial contact between the coating surface and the Al 6061 surface and secondly, solid-state diffusion along the coating/MMC contact interface. The initial contact area between the one side of the coating and the metal surfaces is only a small fraction of the theoretical area available, due to the presences of micro-asperities of the surface of both the metal sample and the coating. However, under the effects of heating and an external pressure an intimate contact can be establish at the bonding surfaces, as the micro-asperities suffer plastic deformation. As the temperature increases greater plastic flow is achieved at the interface and the percentage contact area increases. This increase in the contact area results in an increased diffusivity of the solute (Ni) into the base metal. During the heating stage Ni diffuses deep into the Al-MMCs resulting in the formation of complex intermetallic compounds.

Figure 22(a) shows an SEM micrograph of a joint bonded using a 15µm thick Ni-Al_2O_3 coating as the interlayer. The joint was made at a bonding temperature of 600°C for 1 minute and shows three distinct reaction layers at the interlayer/MMC interface. The nano- Al_2O_3 particles that were co-electrodeposited with Ni can are clearly shown in Figure 22.

The composition of the reaction layers was determined quantitatively by energy dispersive spectroscopy (EDS) analysis and is shown in Table 4. The formation of reaction layers (L_2 and L_3) shown in Figure 22, occurred as a result of the inter-diffusion of Ni and Al. EDS analysis showed that L_1 was composed of a Ni-Al layer dispersed with nano-sized Al_2O_3 particles after a bonding time of 1 minute (see Table 4). Reaction layer L_2 on the other hand appears to be a nickel-aluminide with compositions of 50.7 (at %) Ni and 46.70(at %) Al. The L_3 layer contains approximately 24.30 (at %) Ni and 77.0 (at %) Al, which is likely to be $NiAl_3$ intermetallic. This compound is believed to form due to the low solubility of Ni in Al. This has been reported to be approximately 2.9 at% [46]. The saturation of the aluminum interface through the inter-diffusion of Ni and Al leads to the precipitation of the nickel aluminide intermetallic $NiAl_3$.

The phase diagram of the Ni + Al system indicates the thermodynamic stability of the γ'-Ni_3Al phase when formed in the nickel concentration range of about (74 to 76) at.% [46]. Additionally, the phase diagram of the (Ni + Al) system proposed by Nash et al. [47] showed that for aluminum concentrations exceeding 40 mol%, there exist three coexistence fields: (Al + $NiAl_3$), ($NiAl_3$ + Ni_2Al_3) and the non-stoichiometric intermetallic β -NiAl, which is formed in the concentration range 43mol% to 59 mol% aluminum. Rog et al. [48] determined the Gibbs free energy of formation for various intermetallic compounds forming in the Ni + Al system. The results showed that within the temperature range of 570°C to 620°C (843K to 893K) the nickel aluminide compounds listed in Table 4 are formed.

Based on the scientific literature, the compound formed in the reaction layer L_3 is likely to be the $NiAl_3$. This compound also appears on the right side of Equation 9 and is believed to form due to low solubility of Ni in Al which has been reported to be approximately 2.9 at% [46]. The saturation of the aluminum interface through the diffusion of Ni can lead to the precipitation of the nickel aluminide intermetallic $NiAl_3$. The composition of L_2 indicates that the compound is likely NiAl.

Phase	ΔG_f^0 (kJ.mol^{-1})
Ni +Al = NiAl	-133.0 +/- 1.0
Ni+NiAl$_3$= Ni$_2$Al$_3$	-144.1+/- 0.8
2Ni +3Al = Ni$_2$Al$_3$	-311.0+/- 1.7
Ni + 3Al =NiAl$_3$	-166.8+/- 0.9

Table 4. The standard Gibbs free energy values for the chemical reactions with nickel aluminides at 870K [48]

Layers	Al / wt%	Ni / wt%	Si / wt%	Mg / wt%	Compound
L$_1$	12.5	87.5	0	0	Ni +Al$_2$O$_3$
L$_2$	46.71	50.67	0.27	0	NiAl
L$_3$	72.84	24.31	0	0.55	NiAl$_3$
L$_4$	72.84	24.31	0	0.55	NiAl3

Table 5. Energy dispersive spectroscopic compositional analyses of the reaction layers developed during bonding (wt%)

Figure 22. (a) SEM micrograph of joint bonded with a 15 μm Ni-Al$_2$O$_3$ coating for 1 min.

6.2. Liquid formation and base-metal dissolution

Immediately following the heating stage, eutectic melting ensues. According to Dmitry et al [50] there are two possible reactions which are capable of producing a liquid at the joint interface within this temperature range (570°C-620°C). Upon reaching a temperature of 565°C, the L_2 compound reacts with L_4 (88.4%Al and 2.08 % Si) to form a eutectic liquid (E1) along the bond interface as predicted by Equation 9 [46]. The compound formed in L_2 is consumed in the eutectic reaction and diffuses into the base metal. When the joint region was heated to 577°C a second eutectic liquid formed within L_4 as predicted by Equation 10 [49]. The formation of a eutectic liquid layer at the bond interface leads to faster inter-diffusion between Al and the Ni interlayer and this results in a gradual change in the composition of the joint region.

$$L_1 \xrightarrow{\;565\,^{\circ}C\;} (87\%)\,Al + (2\%)\,Si + (11\%)\,NiAl_3 \tag{9}$$

$$L_2 \xrightarrow{\;577\,^{\circ}C\;} (87.5\%)\,Al + (12.5\%)\,Si \tag{10}$$

Dmitry et al. [50] carried out thermodynamic calculations of (Al-Mg-Si-Fe-Ni) quinary systems formed in aluminum alloys. The results showed that in alloys containing Al-Mg-Si-Fe-Ni, numerous ternary and quaternary reactions can occur that has the potential of producing a liquid phase. Some of the phases that contribute to these reactions are shown in Table 6. The XRD analysis of the fractured surfaces shown in the previous also indicated that presence of $AlFe_3Si$. This phase is possibly a variant of the β-phase family listed in Table 6.

Phase	Designation	Composition (wt%)	Density (g/cm³)
Al_3Ni	ε	42Ni	3.95
Mg_2Si	M	63.2Mg; 36.8Si	4.34
Al_9FeNi	T	4.5-14Fe; 18-28Ni	3.40
Al_5FeSi	β	25-30Fe, 12-15 Si	3.45
$Al_8FeMg_8Si_6$	π	10.9Fe; 14.1Mg; 32.9Si	2.82

Table 6. Chemical composition and density of phases formed in Al-Mg-Si-Fe-Ni system [50]

6.3. Base metal dissolution

Base metal dissolution is also an important part of the second stage of bonding. Research work published [1,2] on the mechanisms of TLP bonding showed that dissolution of the interlayer and base metal occurs simultaneously at the bonding temperature. In this study, holding at a bonding temperature above 577°C resulted in increased diffusion of Ni into the base metal and causes the liquid phase to spread between the bonding surfaces due to the effects of pressure and capillary action. The application of pressure enhances spreading of the liquid phase between the bonding surfaces due to capillary action. This spreading increases the contact area and induces the diffusion of Ni and Al into the liquid phase. This results in more eutectic liquid formation and an increase in the width of the liquid phase at the joint, due to dissolution of a section of the base metal. Further increase in the bonding time to 5 minutes, resulted in the diffusion of the Ni into the base metal and away from the joint interface leading to the formation of eutectic or peritectic liquid along the grain boundary as indicated in the EDS analysis of the eutectic microstructure shown in Figure 23.

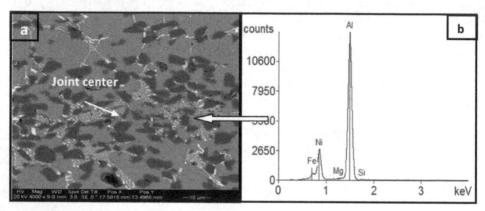

Figure 23. (a) SEM micrograph of joint bonded using a 25 μm Ni foil for 5 min and (b) EDS analysis of the joint region.

According to published studies [20,1] on the stages of TLP bonding, it is expected that dissolution of the interlayer and base metal occurs simultaneously at the bonding temperature followed by spreading of the liquid phase between the bonding surfaces. This spreading increased the bonded area and enhanced the diffusion of Ni and Al into the liquid phase. This continuing diffusion resulted in more liquid formation and an increase in the width of liquid phase. The maximum liquid width attained when using the Ni-Al$_2$O$_3$ composite coating as the interlayer was found by Cooke et al [24] to be $W_{max} = 20.6w_0$.

6.4. Isothermal solidification

In TLP bonding, prolonged hold time at the bonding temperature for 10 minutes allowed for the diffusion of Al into the eutectic liquid which caused the composition of the liquid phase to become aluminum rich, resulting in a change in the eutectic composition (see Table 6). The change in the joint composition initiates the isothermal solidification stage of TLP bonding as a function of bonding time since the temperature and interlayer thickness is constant. When the bonding time is increased to 10 minutes the interface is eliminated and the grain size within the joint increased. This disrupts the band of segregated particles at the interface and homogenized the joint zone. When the bonding time was increased to 30 minutes a corresponding increase was seen in grain size, resulting in more uniform distribution of micro-Al$_2$O$_3$ particles. In the reported studies on transient liquid phase diffusion bonding of Al-MMCs, it was shown that the width of the segregated zone at the joint center increased with increasing bonding time. The opposite of this relationship was seen when using the Ni-Al$_2$O$_3$ coating. As the bonding time increased, the width of the segregated region decrease. This can be attributed to the heterogeneous nucleation of grains within the joint zone during solidification and this lead to grain refining at the joint. The high resolution SEM micrograph shown in Figure 33 revealed the presence of a nano-Al_2O_3 particle at the center of a grain. EDX spectra of the particle showed Al and O in high concentrations with traces of Mg. Comparing Gibbs free energy of formation at the bonding temperature for MgO (-1195 kJ/mol) and Al_2O_3 (-985 kJ/mol) it is found that Al_2O_3 is unstable in the presence of Mg hence it is like that some of the nano-size Al_2O_3 will decomposed to form $MgAl_2O_4$ compound.

The segregation of particles during isothermal solidification was accredited to the pushing of strengthening particles by the solidifying liquid-solid interface. Stefanescu [17, 18] showed that particle pushing can be assumed to be a steady-state condition under which the interface velocity can be assumed to be equal the rate of isothermal solidification This rate can be calculated using a model proposed by Sinclair [51]. The constant, ξ, signifies the solidification rate of the system. Increasing ξ results in faster solid-liquid interface motion, and a shorter duration of the isothermal solidification stage. The rate of isothermal solidification can be calculated using Equation 11.

$$\xi = -2\,(k\ -\ 1)^{-1}\sqrt{\frac{D}{\pi}}\cdot\frac{\exp\left(\dfrac{-\xi^2}{4.D}\right)}{erfc\left(\dfrac{\xi}{2\sqrt{D}}\right)} \tag{11}$$

Where k is a partition coefficient given by $\dfrac{C_{L\alpha}}{C_{\alpha L}}$ and D is the diffusivity of Ni into Al. The diffusivity at 570°C is 4.69 x10^{-13} m²/s, when the bonding temperature is increased to 620°C, the diffusivity also increased to 1.58 x10^{-12} m²/s. This increase in diffusivity is reflected in a faster solid-liquid interface rate (ξ) and a shorter isothermal solidification stage.

The final concentration of Ni $\left(C_{L\alpha}\right)$ was taken from the Al-Ni-Si phase diagram [30] to be 4.9 wt % for the bonding temperature of 620°C. The diffusivity of Ni in Al at 620°C is D = 1.58 x 10^{-12} m²/s [25, 31]. Using these values the predicted interface rate constant ξ = -0.395μm/s was calculated from Equation 8. This solidification rate is significantly less than the critical interface velocity (16 -400 μm/s) required to engulf dispersed particle during solidification [26]. Li et al. [21] suggested that particle segregation tendency is dependent on the relationship between the liquid film width produced at the bonding temperature, particle diameter and inter-particle spacing. When the liquid film width is large enough that sufficient particulate material is contained in the melt, particles will be pushed ahead of the solidifying liquid-solid interface resulting in particle segregation at the bond-line. However if the liquid film width is less than some critical value, segregation should not occur. WDS analysis across the joint zone as a function of bonding time indicated that the Ni volume at the joint center varied between 3.67 wt% after 1 minute and 0.15 vol.% after 30 minutes bonding time. The relationship between the width of segregated region and the maximum liquid width that formed during bonding was determined by using equation 12.

$$S_{sz} = W_{max} \left[\frac{\delta_p + \chi_2}{\delta_p + \chi_1} \right] \qquad (12)$$

Where δ_p is the average particle size, χ_1 is the inter-particle spacing in the as received MMC and χ_2 is the inter-particle spacing after bonding at the joint. By substituting δ_p = 28 μm, χ_1 = 10 μm and χ_2 = 0 into Equation 6.3, the relationship between the width of the particle segregated zone and the maximum width of the eutectic liquid phase was found to be, S_{sz} = 0.74 W_{max}. This means that the width of the segregated zone is approximately 74% of the width of the maximum width of the eutectic liquid phase formed during bonding.

7. Conclusion

Transient liquid phase diffusion bonding of particle reinforced Al-6061 MMC using Ni-Al₂O₃ interlayer was successfully achieved using nano-composite Ni-Al₂O₃ coating. The results obtained from the spectroscopic analyses using WDS and XRD showed that Ni-Al₂O₃ coating the bonding process can be characterized in four distinct stages: interfacial contact and solid-state diffusion, which resulted in the formation of three reaction layers promoted by the diffusion of nickel into the aluminum base metal. The second stage of the joining process was the formation of an Al-Ni-Si eutectic liquid at the bonding temperature. It is supported that the reaction layers formed within the joint melted to form a liquid phase, followed by dissolution of the base metal (third stage) as the liquid spread between the bonding surfaces through capillary action. The final stage of bonding involved isothermal solidification at the bonding temperature in which the diffusion of Ni into Al results in a change in the composition of the liquid phase leading to solidification.

The joint shear strength was studied as a function of bonding parameters, bonding time, bonding temperature, interlayer thickness and interlayer particle size. The results showed that the joint shear strength increased with increasing bonding time, bonding temperature and interlayer thickness. On the other hand the results showed that bond strength increased when the interlayer particle size was reduced from 500 nm to 50 nm. The increase joint shear strength seen when Ni-Al₂O₃ coating were used was attributed to the presence of highly-dispersed nano-sized reinforcement particles in the joint region act to strengthen the joint region by Orowan bowing mechanism.

The results showed that an optimum joint strength of 144 MPa can be achieved if the following bonding parameters are used: 30 minutes bonding time, 620°C bonding temperature, and 11 μm thick Ni-Al₂O₃ coating. Within the parameter ranges tested the bonding pressure had the lease effect on the joint shear strength of TLP bonded joints.

Author details

Kavian Cooke

Department of Mechanical Engineering, School of Engineering, University of Technology, Jamaica

8. References

[1] W. D. Macdonald and T. W. Eager 'Transient Liquid Phase bonding processes' Minerals Metals and Joining Society, 1992, 93-101

[2] I. Tuah-Poku, M. Dollar and T. B. Massalski, 'A study of the transient Liquid phase bonding process applied to a Ag/Cu/Ag sandwich joint' Metall. Trans. A, 1988, 19 A, 675 -686

[3] Y. Zhou, W.F. Gale, and T.H. North Modelling of Transient Liquid. Phase Bonding, International Material Review, (1995) Vol. 40, No.5, pp 181-196

[4] ASTM E3 -01 (2008) Standard preparation for metallographic examination, ASTMInternational 100 Barr Harbor drive United States

[5] Z. Li, Y. Zhou, T.H. North,'Counteraction of particulate segregation during transient liquid-phase bonding of aluminum-based MMC material' Mater. Sci. 1995, 30, 1075–1082

[6] A.A. Shirzadi, E.R. Wallach, (1997) New approaches for transient liquid phase diffusion bonding of Aluminum based metal matrix composites, Material Science Technology Vol.13, pg: 135–142.

[7] Y. Z hou W. F. Gale, and T. H. North, Modeling of transient liquid phase bonding, International Mater. Rev., 1995, Vol. 40, No.5, 181-196.

[8] N. S. Bosco, F. W. Zok, Critical interlayer thickness for transient liquid phase bonding in the Cu–Sn system, Acta Materialia, 2004, Vol. 52. pg: 2965–2972.

[9] W. D. Macdonald and T.W. Eager, 'Transient liquid phase bonding processes' Minerals Metals and Joining Society, 1992, 93-101.

[10] J.R. Askew, J.F. Wilde and T.I. Khan, Transient liquid phase bonding of 2124 aluminum metal matrix composite, Mater. Sci. Tech. 1998, 14 920–924.

[11] R.F. Chen, Y.H. Zhao, Z.X. Shen, L.G. Dai, X.L. Zhang, R. Zhu, Study on the joint strength of SiCp/ Al metal-matrix composite by magnetron sputtering method, Mater. Sci. 2009, pp 628 - 629

[12] A. Suzumura, and Y. Xing, Diffusion brazing of short Al2O3 fiber reinforced aluminum composite. Mater. Trans., 1996, Vol. 37, pg: 1109–1115.

[13] T. Enjo, K. Ikeuchi, Y. Murakami, N. Suzuki, Diffusion bonding of aluminum-magnesium-silicon series 6063 alloy reinforced with alumina short fibers, Trans. of JWRI, 1987, Vol.16, No. 2, pg: 285-92

[14] Y. Zhai, T.H. North and J. Serrato-Rrodrigues, Transient liquid-phase bonding of alumina and metal matrix composite base materials, Mater. Sc., 1997, Vol.32, pg;1393–1397.

[15] A. A. Shirzadi, E.R. Wallach, New approaches for transient liquid phase diffusion bonding of Aluminum based metal matrix composites, Mater. Sci. Tech., 1997, Vol.13, pg: 135–142.

[16] A. Urena, J.M. Gomez de Salazar, and M.D. Escalera, Diffusion bonding of and aluminum metal matrix composites Weld. J, 1997, Vol.76, Pg: 92–102.

[17] D.M Stefanescu, Science and engineering of casting solidification, Kluwer Academic/Plenum Publishers, New York (2002).

[18] D.M. Stefanescu, F.R. Juretzko, B.K. Dhindaw, A. Catalina, S. Sen, P.A. Curreri, "Particle Engulfment and Pushing by Solidifying Interfaces: Part II, Microgravity Experiments and Theoretical Analysis" Metall. Mater. Trans. A, Vol. 29 (1998), pp; 1697.

[19] Z.W. Xu, J.C. Yan, G.H. Wu, X.L. Kong, S.Q. Yang, Interface structure and strength of ultrasonic vibration liquid-phase bonded joints of Al2O3p/6061Al composites Scripta Materialia, 2005, Vol. 53, pg: 835–839

[20] J.H. Huang, Y.L. Dong, Y. Wan, J.G. Zhang, G.A. Zhou, Reactive diffusion bonding of SiCp/Al composites by insert layers of mixed powders, Mater. Sci. Tech., 2005, Vol. 21, pg: 1217–1221

[21] B. Wielage, I. Hoyer, S. Weis, Soldering aluminum matrix composites. Weld J 2007;86:67–70

[22] J. Yan, Z. Xu, L. Shi, X. Ma, S. Yang, Ultrasonic assisted fabrication of particle reinforced bonds joining aluminum metal matrix composites Mater Des. 32 (2011) 343–347

[23] K.O. Cooke, T.I. Khan, G.D. Oliver (2011) Nanostructure particle-reinforced transient liquid phase diffusion bonding; a comparative study, Metall. Mater. Trans. A, DOI: 10.1007/s11661-011-0663-6.

[24] K.O. Cooke, T.I. Khan, G.D. Oliver (2011) Transient liquid phase diffusion bonding Al-6061 using nano-dispersed Ni coatings, Mater. Des. DOI: 10.1016/j.matdes.2011.04.051.

[25] W. Deqing, S. Ziyuan and K. Tangshan, Composite plating of hard chromium on aluminum substrate, Surface and Coatings Technology, 2005, Vol.191, pp:324– 329

[26] D.R. Gabe, Principles of metal surface treatment and protection, Anti-Corrosion Methods and Materials, Published by Pargamon Press, Oxford,1972,

[27] W.F. Gale and D.A. Butts, Transient liquid phase bonding Science and Technology, Journal of Welding and Joining, 2004, Vol. 9, No. 4, pp: 283-300.

[28] W.D. Zhuang and T.W. Eagar, Transient liquid-phase bonding using coated metal powders Welding Journal, 1997, Vol. 76, Issue 12, Pages 157-167.

[29] M.L. Kuntz, Y. Zhou, and S.F. Corbin, A study of transient liquid-phase bonding of Ag-Cu using differential scanning calorimetry, Metallurgical and Materials transaction A, Vol. 37A, 2006, pp: 2493-2504.

[30] D.S. Duvall, W.A Owczarski, and D.F. Paulonis, TLP bonding: a new method of joining heat resisting alloys, Welding journal, Vol 53, No4. 1974, pp: 203-214

[31] W. F. Gale. and T. C. Totemeier 'Smithells Metals Reference Book', 8th Edn, 2004, Published by Elsevier

[32] K. Cooke, T. Khan and G. Oliver (2012) Critical interlayer thickness required for transient liquid phase bonding using nano-dispersed Ni coatings, submitted to Journal of Science and Technology of Welding and Joining Vol. 17 No. 1 page 22-31.

[33] K.O. Cooke, A Study of the Effect of Nanosized Particles on Transient Liquid Phase Diffusion Bonding Al6061 Metal–Matrix Composite (MMC) Using Ni/Al2O3 Nanocomposite Interlayer , Metallurgical and Materials Transaction B, (2012) (accepted article) DOI: 10.1007/s11663-012-9643-5

[34] Y.-L. Shen, E. Fishencord, N. Chawla, Correlating macrohardness and tensile behavior in discontinuously reinforced metal matrix composites Scripta. Materialia, Vol. 42, 2000, pp 427.

[35] Z.Y. Ma, Y.L. Li, Y. Liang, F. Zheng, J. Bi, S.C. Tjong, Nanometric Si3N4 particulate-reinforced aluminum composite, Journal of Material Science and Engineering A 219 (1996) 229.

[36] I. Shao, P.M. Vereecken, C.L. Chien, P.C. Searson and R.C. Cammarata, Synthesis and characterization of particle-reinforced J Mater Res 17 (2002), pp. 1412–1418.

[37] L. Thilly, M. Véron , O. Ludwig, F. Lecouturier Deformation mechanism in high strength Cu/Nb nano-composites Materials Science and Engineering A309–310 (2001) 510–513

[38] S.V. Kamat and M. Manoharan, Work hardening behaviour of alumina particulate reinforced 2024 aluminum alloy matrix composites, Journal of Composite Materials 1993, 27: 1714-1721 DOI: 10.1177/002199839302701801

[39] H. Nami, A. Halvaee and H. Adgi, Transient liquid phase diffusion bonding of Al/Mg2Si metal matrix composite, Materials and Design, 2011, doi:10.1016/j.matdes.2011.02.003

[40] G.A. Chadwick and P.J. Heath, Machining metal matrix composites Mater. 6/2 (1990), pp. 73–76.

[41] Z.Y. Ma, SC. Tjong, In situ ceramic particle-reinforced aluminum matrix composites fabricated by reaction pressing in the TiO2 (Ti)–Al–B (B2O3) systems Metallurgical Material Transaction 1997; 28(A):1931–42.

[42] S. C. Tjong, Novel nanoparticle-reinforced metal matrix composites with enhanced mechanical properties, Advanced Engineering Materials Volume 9, Issue 8, 2007

[43] S. F. Hassan, M. Gupta, "Development of high performance magnesium nano-composites using solidification processing route Journal of Material Science and Technology, 2004, 20, 1383.

[44] N. Srikanth, S. F. Hassan, M. Gupta, Energy dissipation studies of Mg-based nano-composites using an innovative circle-fit approach, Journal of Composite Material.2004, 38, 2037.

[45] Z. Zhang, D. L. Chen, Consideration of Orowan strengthening effect in particulate-reinforced metal matrix nano-composites: A model for predicting their yield strengthScr. Mater. 2006, 54, 1321.

[46] K. W. Richter, K. Chandrasekaran and H. Ipser, The Al–Ni–Si phase diagram. Part II: phase equilibria between 33.3 and 66.7 at.% Ni, Journal of Intermetallics, 12 (2004) 545–554

[47] P. Nash, M.F. Singleton, J.L. Murray, in: P. Nash (Ed.), Phase Diagrams of Binary Nickel Alloys, ASM International, Materials Park, OH, 1991, pp. 3–11.

[48] G. Rog , G. Borchardt, M. Wellen, W. Lose, determination of the activities in the (Ni + Al) alloys in the temperature range 870K to 920K by a solid-state galvanic cell using a CaF2 electrolyte J. Chem. Thermodynamics 35 (2003) 261–268

[49] V. Raghavan, Al-Ni-Si (Aluminum-Nickel-Silicon) Phase diagram evaluations, Journal of Phase Equilibria and Diffusion, 2005, Vol. 26, pp:262-267

[50] E. Dmitry, G. Belov, N.A. Aksenov, and A. Andrey, (2005) Multicomponent Phase Diagrams : Applications for Commercial Aluminum Publisher: Elsevier Science

[51] C.W. Sinclair, 'Modelling transient liquid phase bonding in multi-component systems Journal of Phase equilibria, 1998, 20, No. 4,

Superparamagnetic Behaviour and Induced Ferrimagnetism of LaFeO₃ Nanoparticles Prepared by a Hot-Soap Technique

Tatsuo Fujii, Ikko Matsusue and Jun Takada

Additional information is available at the end of the chapter

1. Introduction

Lanthanum orthoferrite, $LaFeO_3$, is one of the most common perovskite-type oxides having an orthorhombic perovskite structure (space group Pbnm), where the distortion from the ideal cubic structure occurs to form the tilting of the FeO_6 octahedra. $LaFeO_3$ has much practical interest for electroceramic applications due to their attractive mixed conductivity displaying ionic and electronic defects [1, 2]. The mixed ionic-electronic conductivity of $LaFeO_3$ exhibits a linear response to oxygen pressure and provides oxygen sensor applications [3]. The excellent sensitivity and selectivity towards various toxic gases such as CO and NO_x are observed as well [4]. Moreover, $LaFeO_3$ nanoparticles exhibited good photocatalytic properties such as water decomposition and dye degradation under visible light irradiation [5, 6]. These properties are enhanced by the homogeneity and high surface area of the fabricated $LaFeO_3$ particles. Fine particles with diameter of less than 100 nm are potentially required for these purposes.

Besides, the orthoferrites are known to be prototype materials for magnetic bubble devices because of their large magnetic anisotropy with small magnetization [7]. $LaFeO_3$ is an interesting model system of orthoferrite antiferromagnets showing a weak ferromagnetism. The Néel temperature, T_N, of $LaFeO_3$ is 738 K, which is the highest temperature in the orthoferrite family [8]. The magnetic moments of Fe^{3+} ions are aligned antiferromagnetically along the orthorhombic a-axis. But they are slightly canted with respect to one another due to the presence of Dzyaloshinskii-Moriya interaction. A weak ferromagnetic component parallel to the c-axis appears. The magnetization of $LaFeO_3$ bulk crystals is considerably small, 0.044 μ_B/Fe [8]. However, magnetic structures of small particles are often different from those of bulk ones. For instance, antiferromagnetic nanoparticles exhibit increasing net magnetization due to the presence of uncompensated surface spins [9, 10]. If the ferromagnetic behavior is promoted in $LaFeO_3$, it should provide facile handling of their applications by using magnetic field. Magnetic properties of well-defined $LaFeO_3$ nanoparticles are worthy to investigate.

It is well known that the wet-chemical methods offer large advantages for low-temperature oxide formation with high surface area, small particle size, and exact cation-stoichiometry. Several methods such as co-precipitation technique [11, 12], polymerized complex method [13], combustion synthesis [14], and sol-gel technique [15] were reported to prepare $LaFeO_3$ nanoparticles. For instance the formation of a single phase of $LaFeO_3$ with the perovskite structure was observed at lower calcination temperatures of 300°C in [11, 12]. This temperature was much lower than that of conventional solid state reaction method. Recently we have successfully prepared $LaFeO_3$ nanoparticles by using the new chemical synthesis method, so-called "hot soap method" [16, 17]. It showed high controllability over the formation of nanoparticles with narrow size distribution, which was performed in the presence of surfactant molecules at high temperatures. The hot soap method is based on the thermal decomposition of reaction precursors of organometallic compounds in polyol solvent. But the presence of surfactant molecules in the solution prevents aggregation of precursors during growth. It was widely applied to prepare nanoparticles of compound semiconductors [18] and metallic alloys [19]. However there were few reports on preparing oxide nanoparticles [20].

In this paper we describe the details of our synthesis procedure of $LaFeO_3$ nanoparticles by using the hot soap method. The magnetic properties of the resultant particles were also discussed as a function of the particle sizes.

2. Experiment

$LaFeO_3$ nanoparticles were synthesized by the hot soap method. Their synthesis procedure is outlined in Figure 1. All chemicals used in this experiment were of reagent grade and used without any further purification. Iron acetylacetonate ($Fe(acac)_3$) and lanthanum acetate ($La(ac)_3 \cdot 1.5H_2O$) were preferred as iron and lanthanum sources, respectively, that were soluble in organic solvents such as polyethylene glycol (PEG 400). In a typical synthesis procedure, equal amounts of $Fe(acac)_3$ (1.2 mmol) and $La(ac)_3$ (1.2 mmol) were weighed out accurately and charged into a reaction flask with 20 mL of PEG 400. Coordinating organic protective agents of oleic acid (5 mmol) and oleylamine (5 mmol) were injected into the reaction mixture and the transparent brown solution was observed. Thereafter, the mixture solution was raised to 200°C and maintained for 3 h with stirring. Before cooling down to room temperature, 50 mL of ethanol was added to the reaction mixture, in order to precipitate the particles. The precipitated particles were rinsed with ethanol and dried at 100°C for 1 h. Some of the sample powders were heat-treated in air for 6 h at various temperatures between 300 and 500°C.

Obtained sample powders were characterized by x-ray powder diffraction (XRD) with monochromatic Cu Kα radiation, infrared spectroscopy (IR), and thermogravimetry and differential thermal analysis (TG-DTA). Powder morphologies of the products were observed by scanning electron microscopy (SEM, Hitachi S-4300) at 20 kV and transmission electron microscopy (TEM, Topcon EM-002B) at 200 kV. The BET surface areas were measured by using N_2 absorption at 77 K. The magnetic properties were investigated using a vibrating sample magnetometer with high-sensitivity SQUID sensor (MPMS SQUID-VSM) and conventional transmission Mössbauer spectroscopy with a 925 MBq ^{57}Co/Rh source. The velocity scale of Mössbauer spectra was calibrated with reference to α-Fe.

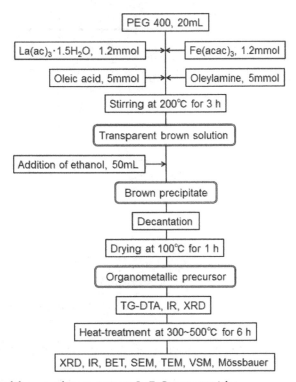

Figure 1. Flowchart of the procedure to prepare LaFeO₃ nanoparticles.

3. Results and discussion

3.1. Thermal decomposition of organometallic precursor

Figure 2 shows the TG-DTA curves of the organometallic precursor obtained by the hot soap method. In the TG curve, there are four temperature regions based on weight loss: (1) RT ~ 220°C, (2) 220 ~ 420°C, (3) 420 ~ 510°C, and (4) 510 ~ 600°C, in which the corresponding organic weight loss of 4%, 38%, 45% and 1% were observed, respectively. The small weight loss of the region (1) was ascribed to the evaporation of residual water and ethanol. While the large weight loss of the region (2), accompanied with faint endo- or exothermal peaks in the DTA curve, corresponded to the sublimation and the decomposition of excessive organic substances such as PEG, oleic acid and oleylamine. The temperature range of the region (2) was coincident with the boiling points of individual substances of 250°C (PEG), 223°C (oleic acid) and 350°C (oleylamine). The region (3) comprised the combustion reaction of the residual organics and carbonate components as suggested by the large exothermal peaks at 460°C and 500°C. The large weight loss was due to the decomposition of the most of the organics by oxidation and the release of NO_x, H_2O, CO and CO_2 gases, together with the formation of $LaFeO_3$ as discussed latter. Further heat-treatment in the region (4) gave no major weight loss anymore.

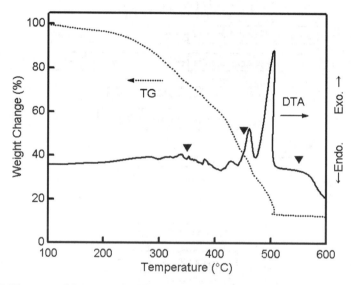

Figure 2. TG-DAT curves of the precursor. The solid triangles indicate the set temperatures for the subsequent XRD and IR observations.

In order to identify the structural changes of the resultant precursor after the heat-treatment, we measured XRD and IR spectra of heat-treated samples taken out from the TG-DTA furnace immediately after reaching to the set temperatures. Figure 3 shows the XRD patterns of the heat-treated samples at various set temperature by TG-DTA. The XRD pattern of the precursor powder had no sharp diffraction lines resulting from the formation of perovskite type oxides. The broad bands centered at around $2\theta = 30°$ and $45°$ suggested the existence of disordered La_2O_3 phase [21]. Scarce changes in XRD patterns were observed up to the heating of 450 °C. However the XRD pattern of the specimen heated at 550 °C showed clear peaks attributed to $LaFeO_3$ perovskite phase. The pattern showed only the presence of the orthorhombic $LaFeO_3$ phase without the broad bands. Observed crystallization temperature between 450 and 550°C was very consistent with the thermal decomposition temperature of the precursor associated with the large exothermal peaks on the corresponding TG-DTA curves (Figure 2).

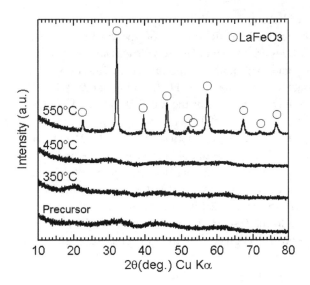

Figure 3. Typical XRD patterns of the obtained products heat-treated at various set temperatures in TG-DTA furnace with no holding time.

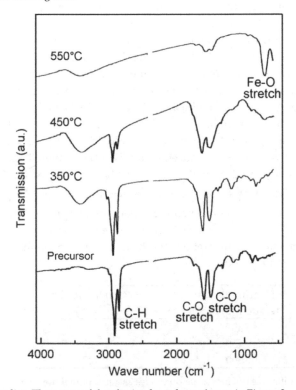

Figure 4. Corresponding IR spectra of the obtained products shown in Figure 3.

The crystallization of LaFeO₃ was fully accompanied with the decomposition of the organic substances. The corresponding IR spectra of the heat-treated samples at various set temperatures of TG-DTA are shown in Figure 4. The IR spectra clearly provided the evidence for the presence of organic substances in the precursor. The strong absorption bands at about 2900 cm^{-1} were assigned to the olefinic C-H stretching bands. While the bands appearing at 1400 ~ 1600 cm^{-1} were characteristics of the metal chelate compounds with carbonyl groups [22, 23]. Two strong peaks at 1550 and 1450 cm^{-1} were mainly assigned to the C=O asymmetric and symmetric stretching modes, respectively. The metal ions in the precipitate could be coordinated with large amount of organic molecules and the large weight loss (87%) at 500°C was confirmed after the thermal decomposition. With increasing the heat-treated temperature, the absorption bands assigned to the organic bonds were gradually decreased in intensity and nearly vanished at 550°C. Besides the strong new band appeared at 600 cm^{-1}, that can be attributed to the Fe-O stretching vibration band due to the formation of LaFeO₃ phase [24]. These results fully agreed with the XRD phase-analysis findings.

3.2. Formation of LaFeO₃ nanoparticles

As shown in Figure 2 in the previous section, a large weight loss of the precipitate started at 220°C, much lower than the temperature of the combustion of the residual organics and the crystallization of LaFeO₃. Moderate decomposition of organic substances in the precursor seemed to start prior to the combustion. The migration of metal ions should be facilitated even at the lower temperature. Therefore the long time heat-treatments below the combustion temperature were conducted to the precursors to prepare nanocrystalline particles. Figure 5 shows the XRD patterns of the samples heat-treated in air for 6 h at various temperatures below the combustion. The sample heated at 325°C showed the diffuse XRD pattern with no crystalline phases. While at 350°C, broad XRD peaks attributed to the LaFeO₃ perovskite phase started to be observed. This temperature was much lower than the crystallization temperature deduced by the TG-DTA curves. The average grain size, t, of LaFeO₃ particles estimated from the XRD peak broadening for the LaFeO₃(121) lines was summarized in Table 1 by using the Scherrer equation: $t = \frac{0.9\lambda}{Bcos\theta}$, where λ is the x-ray wavelength, B is the line broadening at half the maximum intensity (FWHM) in radians, and θ is the Bragg angle. The particle size of LaFeO₃ formed at 350°C was about 16 nm. With increasing the heat-treatment temperature, the XRD peaks became sharper because of the grain growth of the LaFeO₃ particles.

Heat-treatment temperature (°C)	Crystallite size (nm)
350	16
450	20
500	22

Table 1. Average crystallite size of LaFeO₃ nanoparticles as a function of the heat- treatment temperature.

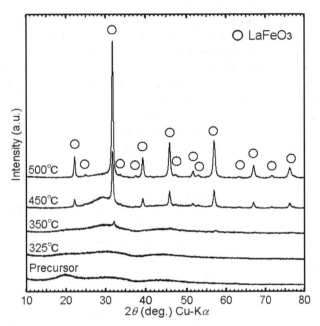

Figure 5. Typical XRD patterns of the obtained products after the long time heat-treatment (6 h) at various temperatures.

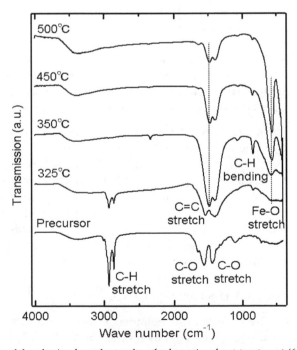

Figure 6. IR spectra of the obtained products after the long time heat-treatment (6 h) shown in Figure 5.

Decomposition of organic substances after the heat-treatment below the combustion temperature was examined by IR spectroscopy. Figure 6 shows the IR spectra of long time (6h) heat-treated samples at various temperatures. The precursor should consist of organometallic composites derived from starting materials. The IR spectrum indicated the clear bands assigned to C-H stretching (~2900 cm^{-1}) and C-O stretching (1400~1600 cm^{-1}) of organic substances [22, 23]. With increasing the heat-treatment temperature, the intensities of initial bands steeply decreased and a new band assigned to Fe-O stretching (~600 cm^{-1}) started to appear at 325°C. This temperature was slightly lower than the formation temperature of LaFeO$_3$ nanoparticles confirmed by XRD observations. And the Fe-O stretching band was intensified monotonically with increasing the temperature. By the way, subsequent to diminishing the intensity of olefinic C-H stretching bands, new strong band appeared in the spectra at 1475 cm^{-1} in addition to a weak band at 843 cm^{-1}. The former band could be assigned to aromatic C=C stretching vibration, while the latter was to aromatic C–H bending one [25]. This result suggested that the framework of organic matrix in precursor was considerably changed after the long time heat-treatment below the combustion temperature. The heat-treatment seemed to promote carbonization of organic substances as well as the crystallization of LaFeO$_3$ nanoparticles.

SEM images of the obtained powders heat-treated in air at various temperatures for 6 h are shown in Figure 7. The precursor particles with and without heat-treatment at 325°C revealed the smooth surface with squamous and wrinkled structures, respectively. These surface morphologies should be very consistent with the amorphous nature of the resultant organic matrix. With increasing the heat-treatment temperature above 350°C, the particles were crowned by squamous surfaces. The macrosized particles seemed to be disaggregated into fine particles after the crystallization. In spite of the nanometric crystal size determined from the XRD profiles, the BET surface area of the obtained LaFeO$_3$ particles at 350 and 500°C were only 3.9 and 3.1 m^2/g, respectively. The nanocrystalline fine particles formed in the precursors were strongly agglomerated with each other in present cases. The residual carbon matrix could also affect the morphology of the obtained powders.

The structure of the LaFeO$_3$ nanocrystallites was further analyzed by TEM micrograph. Figure 8 shows the TEM image and electron diffraction (ED) pattern of LaFeO$_3$ nanoparticles obtained by the long time (6 h) heat-treatment at 350°C. The particulates consisted of an agglomeration of numerous spherical primary particles loosely aggregated together, in contrast to the SEM observations. The average diameter of the primary particles estimated from the TEM image was about 15 nm, which was well consistent with that from XRD. The high-contract dotted rings in the ED patterns indicated the good crystallinity of the LaFeO$_3$ nanoparticles. These rings were indexed as the LaFeO$_3$ perovskite phase with random orientation. No diffraction spots attributed to the secondary phase were observed.

3.3. Magnetic properties of LaFeO$_3$ nanoparticles

Room temperature Mössbauer spectra of obtained powders after the long time heat-treatment (6h) at various temperatures are shown in Figure 9. The sample heated at 500°C, which had a larger crystallite size, showed a clear sextet pattern with a small amount of a

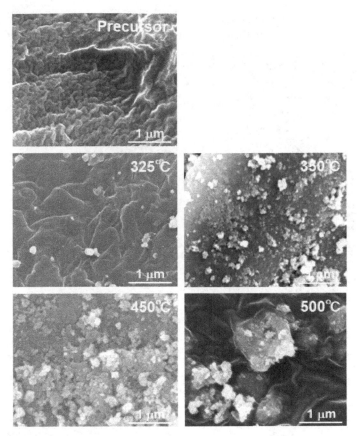

Figure 7. SEM images of the obtained products heat-treated at various temperatures.

doublet component. The observed Mössbauer parameters showing antiferromagnetic ordering were fully consistent with those of the LaFeO₃ bulk crystals [26]. The paramagnetic doublet patterns were dominant for other samples obtained at lower temperatures (350 and 450°C) in spite of the formation of LaFeO₃, as well as the non-crystalline samples at 300 and 325°C. This behavior was attributed to superparamagnetism because of the fine crystallite size of LaFeO₃ as discussed above. By the way, the doublet patterns showed asymmetric profiles. The asymmetric doublets were probably caused by the presence of iron ions in different environments such as surface or inner part of the crystals and/or the non-crystalline matrix. But it was difficult to decompose such complex spectra into unique components. Therefore the spectra were simply analyzed to assume one asymmetric doublet with different widths. The fitted parameters are listed in Table 2. The isomer shift values of ~0.3 mm/s for both doublet and sextet peaks indicated the Fe valence states in the all specimens were trivalent. No reduction occurred during the long time heat-treatment with organic substances. When looking in detail, the Mössbauer parameter of the doublet pattern indicated the systematic change depending on the heat-treatment temperature. The isomer shift gradually decreased while the quadrupole splitting gradually increased with increasing the heat-treatment temperature. This result suggested the formation of strong Fe-O bonds and ligand fields due to the crystallization of LaFeO₃ particles.

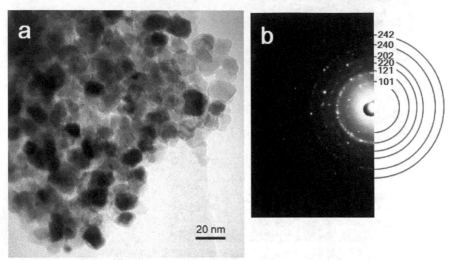

Figure 8. (a) TEM image and (b) its ED pattern of LaFeO₃ nanoparticles obtained at 350°C. Diffraction rings are indexed as their corresponding LaFeO₃ planes.

Heat-treatment temperature (°C)	Isomer shift (mm/s)	Quadupole splitting (mm/s)	Hyperfine filed (kOe)	Intensity (%)
500	0.369	-	517	65
	0.285	1.051	-	35
450	0.364	-	508	23
	0.292	1.008	-	77
350	0.355	-	512	21
	0.313	0.924	-	79
325	0.329	0.891	-	100
300	0.327	0.824	-	100

Table 2. Fitted Mössbauer parameters obtained from the spectra in Figure 9.

Magnetic measurements were performed on the heat-treated powders at various temperatures. The room-temperature magnetization curves of obtained samples are shown in Figure 10 as a function of the heat-treated temperature. LaFeO₃ is known to be antiferromagnetic with weak ferromagnetism [8]. The net magnetization of LaFeO₃ should be very small because of the antiferromagnetic order of the Fe^{3+} spins. Only a slight canting of the adjacent Fe^{3+} spins produced weak ferromagnetism. The spontaneous magnetization of LaFeO₃ bulk crystals was only ~0.1 emu/g at room temperature [27]. However all samples seemed to have large magnetization, especially for the sample heat-treated at 350°C. The shape of hysteresis loops showing the small spontaneous magnetization were characteristics of weak ferromagnetism. The maximum field applied of 60 kOe did not saturate the magnetization. This was mainly caused by antiferromagnetic ordering of the spins in the nanoparticles [27]. By the way, the sample prepared at 500°C showed the remarkable hysteresis gap than the other samples. Mössbauer spectra indicated that only the sample prepared at 500°C possessed predominant antiferromagnetic ordering at room temperature while the others were paramagnetic. The small hysteresis gaps of the samples obtained at lower temperatures (325~450°C) supported their superparamagnetic nature.

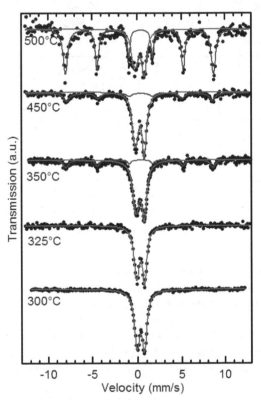

Figure 9. Room temperature Mössbauer spectra of the obtained products heat-treated at various temperatures.

The temperature dependence of magnetization behaviors under zero-field-cooling (ZFC) and field-cooling (FC) conditions measured in an external magnetic field of 200 Oe are shown in Figure 11. Magnetization of LaFeO₃ particles obtained at 500°C was gradually decreased with decreasing the temperature. This behavior was characteristic of the antiferromagnetic ordered states below the T_N. The higher T_N above the room temperature was confirmed by the Mössbauer spectroscopy. The difference of magnetization values between ZFC and FC curves was attributed to the residual paramagnetic components at room temperature. On the other hand, magnetization of other samples obtained at 300~450°C were smoothly increased

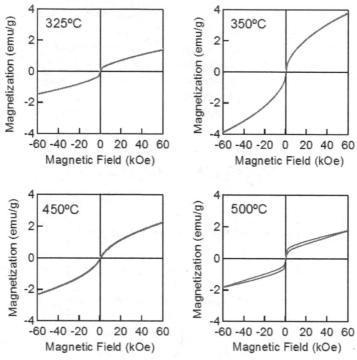

Figure 10. Room temperature magnetization curves of the obtained products heat-treated at various temperatures.

with the decreasing temperature. These behaviors were very consistent with the paramagnetic nature of the obtained samples at room temperature. Moreover, the sharp cusps in the ZFC curves were observed for the samples obtained at 325~450°C. These were considered to be the superparamagnetic blocking temperature (T_B), where the magnetization of LaFeO$_3$ nanoparticles begun to freeze [28]. The T_B detected by the ZFC curves was summarized in Table 3. It was decreased steeply with decreasing the heat-treatment temperature coincident with the decreasing particle sizes. Besides, the observed T_B values were very small, which indicated that the magnetic interactions between the obtained LaFeO$_3$ nanoparticles were very weak. The surface of LaFeO$_3$ nanoparticles was probably coated with the residual organic molecules to form the non-magnetic layers [29]. This assumption was fully supported by the IR spectra indicating the presence of carbonized components and the TEM micrograph showing the loosely aggregating LaFeO$_3$ nanoparticles. It should be mentioned here that the sample heat-treated at 325°C indicated the superparamagnetism, showing the small cusp in the ZFC curve at 7.7 K. The sample seemed to contain the LaFeO$_3$ nanoclusters, though it showed the non-crystalline XRD pattern. But the IR spectrum seemed to have good sensitivity to detect their formation as well as the magnetization measurement. By the way, in the case of conventional superparamagnetic behavior, the FC magnetization curves should increase continuously with decreasing the temperature below the T_B. But the FC curve of the heat-treated sample at 450°C showed a small cusp on cooling as well as the ZFC curve. This

behavior suggested that the antiferromagnetic interactions between nanoparticles were
present [28]. Long-range antiferromagnetic order was developed in associate with the grain
growth of the LaFeO₃ nanoparticles.

Heat-treatment temperature (°C)	Blocking temperature (K)
325	7.7
350	27.6
450	36.9

Table 3. Blocking temperature (T_B) obtained from the ZFC curves shown in Figure 11.

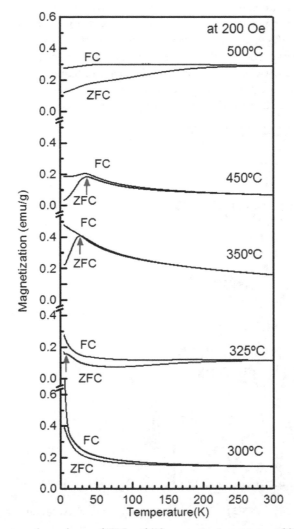

Figure 11. Temperature dependence of ZFC and FC magnetization curves at 200 Oe for the obtained
products at various heat-treatment temperatures

In order to demonstrate the superparamagnetic behavior more directly, we measured the temperature dependence of their Mössbauer spectra down to 15 K. Figure 12(a) shows the typical law temperature Mössbauer spectra of the LaFeO$_3$ nanoparticles obtained at 350°C. The spectrum measured at room temperature (see Figure 9) was well consistent with that of the LaFeO$_3$ nanoparticles with average grain size of 16 nm reported in literature [26]. The spectrum was mainly composed of the super-paramagnetic doublet peak besides to the magnetically split sextet. The component showing the sharp sextet pattern could be attributed to the unexpected coarse LaFeO$_3$ particles which had a good crystallinity, while the most of the particles exhibited the paramagnetic doublet pattern at room temperature. With decreasing the temperature, the paramagnetic spectra corresponded to the enlarged sextets due to the presence of magnetic order. Therefore the spectra were decomposed into three components. One was the sharp sextet for the coarse particles; second and third were the broad sextet and paramagnetic doublet for the nanoparticles, respectively. Relative intensity ratio of each component was plotted in Figure 12(b) as a function of the temperature. The relative area of the paramagnetic component varies almost linearly with decreasing the temperature below 100 K. According to the Mössbauer measurements, we defined the

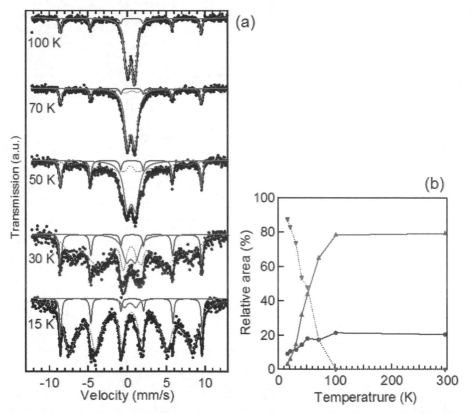

Figure 12. (a) Low temperature Mössbauer spectra of LaFeO$_3$ nanoparticles obtained at 350°C, and (b) relative area of each component as a function of the temperature. The details are described in the text.

median blocking temperature (T$_{Bm}$), as a temperature where the doublet collapsed to 50% of its initial value. This analysis yielded the T$_{Bm}$ of ~50 K, which was much higher than the superparamagnetic T$_B$ obtained by the magnetization measurement. It was well-known that the time scale of the superparamagnetic fluctuations becomes comparable to the time scale of the measurement [30]. The time scale of Mössbauer measurements (~10^{-9} s) was much faster than that of magnetization measurements (~10^{-1} s) and gave the higher T$_{Bm}$ well above the T$_B$. This phenomenon evidenced the presence of superparamagnetic relaxation of the obtained samples.

Figure 13 shows the field dependent magnetization curves measured at lower temperature (5 K) for the LaFeO$_3$ nanoparticles prepared at various heat-treatment temperatures. The samples showed considerable magnetization at low temperature, though the net magnetization of LaFeO$_3$ bulk crystals should be very small because of the antiferromagnetic order of the Fe^{3+} spins. A spontaneous magnetization of the bulk crystals was evaluated to 0.15 emu/g at 5 K [16]. We have recently reported that the LaFeO$_3$ nanoparticles prepared at 350°C revealed an anomalous large spontaneous magnetization of 7.8 emu/g [16]. This phenomenon was almost reproducible. The spontaneous magnetization of the obtained powders exhibited the maximum value for the samples prepared at 350°C. The LaFeO$_3$ nanoparticles with the average diameter of ~15 nm seemed to have large ferromagnetism instead of antiferromagnetism. The induced ferromagnetism in LaFeO$_3$ nanoparticles was possibly associated with the uncompensated surface spins as reported in other antiferromagnetic nanocrystalline systems [9, 10]. It should be worth to mention that the induce magnetic moment per Fe^{3+} ion was evaluated to 0.34 μ$_B$ for the sample obtained at 350°C. Taking into account that the moment of Fe^{3+} is 5 μ$_B$, a large fraction (~7%) of Fe^{3+} ions contribute to the magnetization of LaFeO$_3$ nanoparticles. Magnetic structure of LaFeO$_3$ nanoparticles should be strongly modified from the bulk one.

By the way, all magnetization curves revealed clear hysteresis gaps with small loop shifts at 5 K. The coercivity and the exchange bias field of each sample are summarized in Table 4 as well as the spontaneous magnetization. The exchange field was linearly increased with increasing the heat-treatment temperature. The exchange bias was generally attributed to the exchange coupling between the ferromagnetic shell and the antiferromagnetic core of the particles [31]. In the case of LaFeO$_3$ particles, the uncompensated surface spins and/or the enhanced spin cantings near the surface region should cause the considerable magnetization. With the increasing particle size, the reliable antiferromagnetic core was formed to produce the large exchange coupling.

Heat-treatment temperature (°C)	Spontaneous magnetization (emu/g)	Coercivity (Oe)	Exchange bias (Oe)
325	2.89	823	23
350	5.90	1304	25
450	2.94	2240	64
500	1.40	1275	208

Table 4. Magnetic parameters at 5 K of the obtained products at various heat-treatment temperatures.

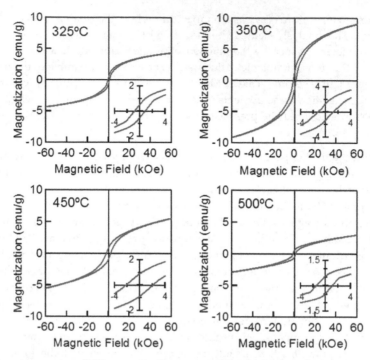

Figure 13. Low temperature (5 K) magnetization curves of the obtained products heat- treated at various temperatures. The insets are the enlarged view of the central region.

4. Conclusion

Nanocrystalline $LaFeO_3$ particles with an average diameter of 16 nm were prepared by using hot-soap technique. Crystalline $LaFeO_3$ were formed at a relatively low heat-treatment temperature of 350°C without any phase segregation. This temperature was much lower than the combustion point of organic substances. With increasing the heat-treatment temperature, the grain size of $LaFeO_3$ particles was increased. The $LaFeO_3$ nanoparticles exhibited superparamagnetic behavior and had anomalous large net magnetization. The spontaneous magnetization reached the maximum when the precursor was heat-treated at 350°C. Moreover the particles exhibited the exchange bias properties with increasing the grain size of $LaFeO_3$. The low blocking temperature observed by both magnetization and Mössbauer measurements indicated that the obtained nanoparticles at 350°C were magnetically isolated with each other.

Author details

Tatsuo Fujii*, Ikko Matsusue and Jun Takada
Department of Applied Chemistry, Okayama University, Japan

---

* Corresponding Author

Acknowledgement

The authors wish to thank their colleagues and students for their technical supports. In particular, the authors would like to acknowledge the contributions of Mr. Takuya Nonoyama for the sample preparations.

5. References

[1] Ming Q, Nersesyan MD, Wagner A, Ritchie J, Richardson JT, Luss D, Jacobson AJ, Yang YL (1999) Combustion synthesis and characterization of Sr and Ga doped LaFeO₃. Solid State Ionics 122: 113-121.

[2] Yoon JW, Grilli ML, Di Bartolomeo E , Polini R, Traversa E (2001) The NO₂ response of solid electrolyte sensors made using nano-sized LaFeO₃ electrodes. Sens Actuators B 76: 483-488.

[3] Hole I, Tybell T, Grepstad JK, Wærnhus I, Grande T, Wiik K (2003) High temperature transport kinetics in heteroepitaxial LaFeO3 thin films. Solid-State Electronics 47: 2279-2282.

[4] Toan NN, Saukko S, Lantto V (2003) Gas sensing with semiconducting perovskite oxide LaFeO₃. Physica B 327: 279–282.

[5] Parida KM, Reddy KH, Martha S, Das DP, Biswal N (2010) Fabrication of nanocrystalline LaFeO₃: An efficient solegel auto-combustion assisted visible light responsive photocatalyst for water decomposition. Int J Hyd Energy 35: 12161-12168.

[6] Su H, Jing L, Shi K, Yao C, Fu H (2010) Synthesis of large surface area LaFeO₃ nanoparticles by SBA-16 template method as high active visible photocatalysts. J Nanopart Res 12: 967-974.

[7] Szymczak R (1977) Temperature dependence of bubble domain structure in YFeO₃ and DyFeO₃ orthoferrites. Physica B+C 86-88: 1351-1353.

[8] Treves D (1965) Studies on Orthoferrites at the Weizmann Institute of Science. J Appl Phys 36: 1033-1039.

[9] Kodama RH, Berkowitz AE (1999) Atomic-scale magnetic modeling of oxide nanoparticles. Phys Rev B 59: 6321-6336.

[10] Lee YC, Parkhomov AB, Krishnan KM (2010) Size-driven magnetic transitions in monodisperse MnO nanocrystals. J Appl Phys 107: 09E124-1-3.

[11] Nakayama S (2001) LaFeO3 perovskite-type oxide prepared by oxide-mixing, co-precipitation and complex synthesis methods. J Mater Sci 36: 5643-5648.

[12] Li X, Zhang H, Zhao M (1994) Preparation of nanocrystalline LaFeO₃ using reverse drop coprecipitation with polyvinyl alcohol as protecting agent. Mater Chem Phys 37: 132-135.

[13] Popa M, Frantti J, Kakihana M (2002) Lanthanum ferrite LaFeO₃₊d nanopowders obtained by the polymerizable complex method. Solid State Ionics 154-155: 437-445.

[14] Wang JB, Liu QF, Xue DS, Li FS (2002) Synthesis and characterization of LaFeO₃ nano particles. J Matter Sci Lett 21: 1059-1062.

[15] Rajendran M, Bhattacharya AK (2006) Nanocrystalline orthoferrite powders: Synthesis and magnetic properties. J Eur Ceramic Soc 26: 3675-3679.

[16] Fujii T, Matsusue I, Nakatsuka D, Nakanishi M, Takada J (2011) Synthesis and anomalous magnetic properties of LaFeO$_3$ nanoparticles by hot soap method. Mater Chem Phys 129: 805-809.

[17] Fujii T, Matsusue I, Nakanishi M, Takada J (2012) Formation and superparamagnetic behaviors of LaFeO$_3$ nanoparticles. Hyperfine Interact 205: 97-100.

[18] Murray CB, Noms DJ, and Bawendi MG (1993) Synthesis and Characterization of Nearly Monodisperse CdE (E = S, Se, Te) Semiconductor Nanocrystallites. J Am Chem Soc 115: 8706-8715.

[19] Sun S, Murray CB, Weller D, Folks L, Moser A (2000) Monodisperse FePt Nanoparticles and Ferromagnetic FePt Nanocrystal Superlattices. Science 287: 1989-1992.

[20] Shouheng Sun S, Zeng H, Robinson DB, Raoux S, Rice PM, Wang SX, and Li G (2004) Monodisperse MFe$_2$O$_4$ (M= Fe, Co, Mn) Nanoparticles. J Am Chem Soc 126: 273-279.

[21] Sadaoka Y, Aono H, Traversa E, Sakamoto M (1998) Thermal evolution of nanosized LaFeO$_3$ powders from a heteronuclear complex, La[Fe(CN)$_6$]·nH$_2$O. J Alloys Comp 278: 135-141.

[22] Tayyari SF, Bakhshi T, Ebrahimi M, Sammelson RE (2009) Structure and vibrational assignment of beryllium acetylacetonate. Spectrochim Acta A 73: 342-347.

[23] Nakamoto K, McCarthy PJ, Martell AE (1961) Infrared Spectra of Metal Chelate Compounds. III. Infrared Spectra of Acetylacetonates of Divalent Metals. J. Am. Chem. Soc. 83: 1272-1276

[24] Gosavi PV, Biniwale RB (2010) Pure phase LaFeO$_3$ perovskite with improved surface area synthesized using different routes and its characterization. Mater Chem Phys 119: 324-329.

[25] El-Hendawy AA (2006) Variation in the FTIR spectra of a biomass under impregnation, carbonization and oxidation conditions. J Anal Appl Pyrolysis 75: 159-166.

[26] Li X, Cui X, Liu X, Jin M, Xiao L, Zhao M (1991) Mössbauer spectroscopic study on nanocrystalline LaFeO$_3$ materials. Hyperfine Interact 69: 851-854.

[27] Shen H, Cheng G, Wu A, Xu J, Zhao J (2009) Combustion synthesis and characterization of nano-crystalline LaFeO$_3$ powder. Phys Status Solidi A 206: 1420-1424.

[28] Bedanta S, Kleemaann W (2009) Supermagnetism. J Phys D Appl Phys 42: 013001-1-27.

[29] Mørup S, Frandsen S, Bødker F, Klausen SN, Lefmann K, LindgÅrd PA, Hansen MF (2002) Magnetic Properties of Nanoparticles of Antiferromagnetic Materials. Hyperfine Interact 144/145: 347-357.

[30] Bødker F, Hansen MF, Koch CB, Lefmann K, Mørup S (2000) Magnetic properties of hematite nanoparticles. Phys Rev B 61: 6826-3838.

[31] Ahmadvand H, Salamati H, Kameli P, Poddar A, Acet M, Zakeri K (2010) Exchange bias in LaFeO$_3$ nanoparticles. J Phys D Appl Phys 43: 245002-1-5.

Organic Spectroscopy

Novel Fischer's Base Analogous of Leuco-TAM and TAM⁺ Dyes – Synthesis and Spectroscopic Characterization

Sam-Rok Keum, So-Young Ma and Se-Jung Roh

Additional information is available at the end of the chapter

1. Introduction

Triarylmethane (TAM) dyes are organic compounds containing triphenylmethane backbones. TAM compounds are sometimes called leuco-TAMs (LTAMs) or leuco-bases. (Nair et al., 2006) LTAM molecules are the precursors of TAM⁺ dyes since TAM⁺ dyes are the oxidized form of LTAM molecules. Backbones of TAM molecules are also known to be an important group in intermediates in the synthesis of various organic functional compounds, including the preparation of polymers and supramolecules. (Bartholome & Klemm, 2006)

TAM⁺ dyes potentially have numerous applications in the chemical, pharmaceutical, and life science industries, including as staining agents, ink dyes, thermal imaging materials, carbonless copying materials, drugs, leather, ceramics, cotton, and as a cytochemical staining agent. (Balko & Allison, 2000) A number of TAM⁺ dye molecules are well known, such as malachite green (MG), brilliant green, crystal violet, and pararosanilin, *etc*. The chemical structures of some of these well-known TAM⁺ dyes are shown in Fig. 1.

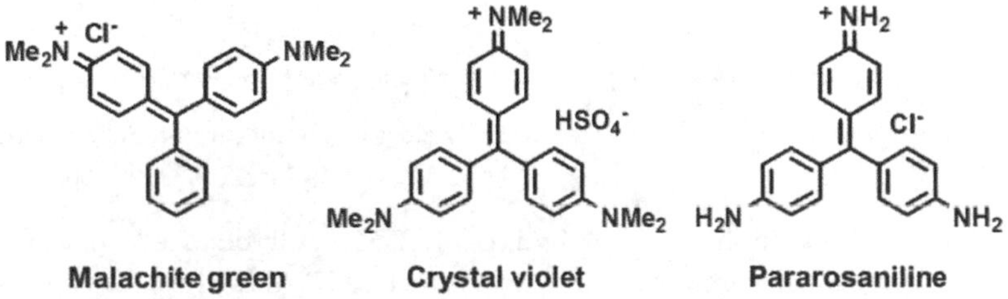

Malachite green   **Crystal violet**   **Pararosaniline**

Figure 1. Chemical structures of well-known TAM⁺ dyes.

Among them, MG is one of the most commonly used chemicals in dye chemistry. MG is a common green dye but it is absorbed into the human body in its carbinol and leuco forms (see the section on UV-Vis spectroscopy). MG is very active with the fungus *Saprolegnia*, which infects fish eggs in commercial aquaculture, and is known to be good for *Ichthyophthirius* in fresh water aquaria. (Indig et al., 2000) It has been known, however, for MG to be highly toxic to mammalian cells, even at low concentrations.(Plakas et al., 1999; Cho et al., 2003) Because of its low cost, effectiveness as an antifungal agent for commercial fish hatcheries, and ready availability, many people can be exposed to this dye through the consumption of treated fish. Since MG is similar in structure to carcinogenic triphenylmethane dyes, it may be a potential human health hazard.

In addition, in their oxidized form, TAM+ dyes are highly absorbing fluorophores with extinction coefficients of ~2.0 × 10⁵ mol⁻¹cm⁻¹ and high quantum yields. Their absorption maxima can easily be matched with the laser lines by simply changing the length of the conjugated chain and/or the heterocyclic moiety. Thus, TAM+ dyes can be employed as fluorescence labels and sensors of biomolecules *in vivo* because their spectra reach the near-infrared region. Özer (2002) reported efficient non-photochemical bleaching of a TAM+ dye by chicken ovalbumin and human serum albumin, showing that dye–protein adducts can also form and suggesting that proteins may be primary, rather than indirect, targets of TAM+ action. However, the use of this substance has been banned in many countries because of its toxicity and possible carcinogenicity. Substitutive materials for MG compounds have thus been in considerable demand. Numbers of researchers (Gessner & Mayer, 2005) have been interested in developing TAM+ molecules. We developed Fischer's base (FB) analogs of LTAM molecules (Keum et al., 2008, 2009, 2010, 2011, 2012), whose chemical structures contain a couple of heterocyclic FB rings and a substituted phenyl group on a central carbon of the molecules. The structures and numbering system for the Fischer's base (FB) analogs of LTAMs are shown in Fig. 2.

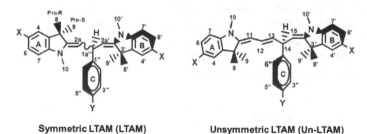

Symmetric LTAM (LTAM)   **Unsymmetric LTAM (Un-LTAM)**

Figure 2. Structures and numbering systems for the FB analogs of symmetric and unsymmetric LTAM molecules.

In this chapter, abbreviations (LTAM and Un-LTAM) will be used to designate the Fischer's base analogs of symmetric and unsymmetric LTAMs, respectively. Note that "the general LTAM" in Section 2.2 denotes the LTAM/TAM+ dyes that contain no FB moieties.

2. Preparation

2.1. General leuco-TAM and TAM⁺ molecules

Generally, the Friedel-Crafts-type catalytic alkylation of aromatic rings with aromatic aldehydes is an effective method for TAM⁺ formation. (Li et al., 2008; Kraus et al., 2008) Several mild and efficient triaryl- and triheteroarylmethanes formations using [Ir(COD)Cl]₂-SnCl₄, AuCl₃, Cu(OTf)₂, and Sc(OTf)₃ as catalysts have also been reported. Grignard reagents or n-butyl lithium compounds have also been used for their preparation. A brief summary of the preparation methods of general LTAM/TAM⁺ molecules is shown in Fig. 3.

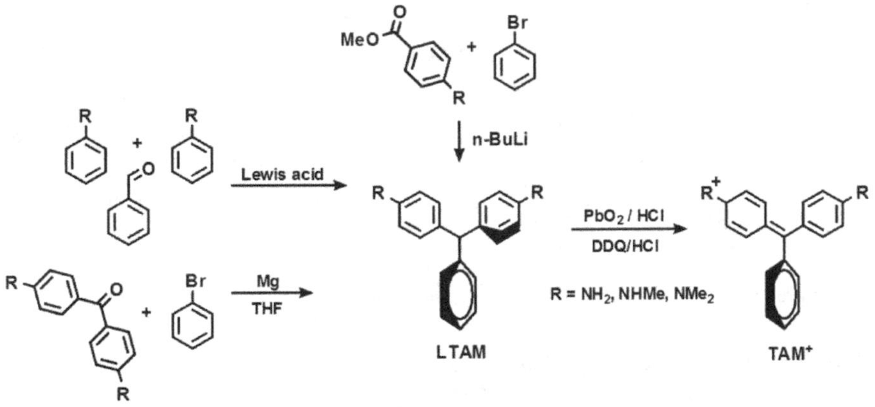

Figure 3. The synthetic procedures for the commercially well-known LTAM molecules.

Although a number of methods are available for the synthesis of triarylmethanes, most are multistep processes and/or require harsh reaction conditions.

2.2. Fischer's base analogs of Leuco-TAM

Fischer's base analogs of LTAM molecules can be obtained from a reaction of a molar excess of Fischer's base and substituted aryl aldehydes. The prepared LTAM dyes consist of two FB rings on the central carbon, where a substituted phenyl ring is located. The LTAM molecules can be symmetric or unsymmetric, depending on the identity of the two FB rings. They are the precursors of the TAM⁺ dyes, which are structurally close to the polymethine dyes (*e.g.*, Cy3, Cy5, *etc.*). (Ernst et al., 1989)

2.2.1. Symmetric LTAM FB analogs

The FB analogs of symmetric leuco-TAM molecules {2,2'-(2-phenylpropane-1,3-diylidene)bis(1,3,3-trimethylindoline)} derivatives were obtained from the reaction of 5-substituted benzaldehyde and excess (2- to 3-fold) FB in ethanol at room temperature for 2–4 h, as shown in Fig. 4. The white precipitate was filtered from the reaction mixture and washed thoroughly with cold ethyl alcohol. Purification was carried out through precipitation from acetone. TAM$^+$ dyes were then obtained from a reaction of LTAM molecule with 2,3-dichloro-5,6-dicyano-1,4-benzoquinone (DDQ) in the presence of HCl, followed by separation of the deep blue form from the product mixtures by column chromatography in MC/MeOH (9:1).

Melting points, yields, and other characteristic data of the prepared LTAM molecules are summarized in Table 1.

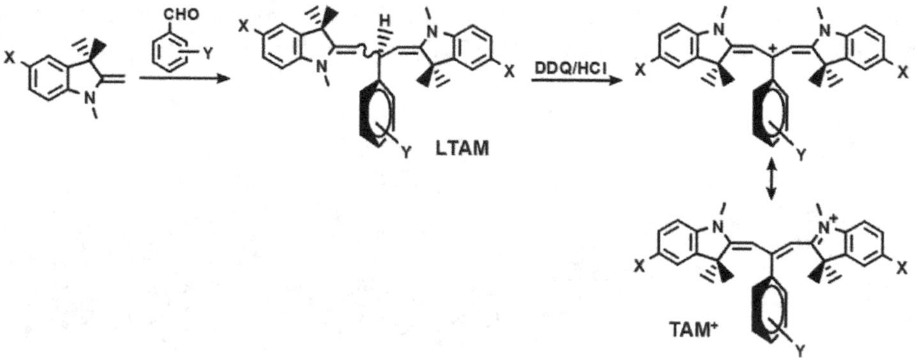

Figure 4. Synthetic scheme for symmetrical LTAM molecules.

2.2.2. Unsymmetric LTAM FB analogs

Unsymmetric LTAMs (Un-LTAM) were obtained from a reaction of excess Fischer's base with the substituted cinnamaldehydes, as shown in Fig. 5. The Un-LTAMs have two different FB skeletons on the central carbon, 1,3,3-trimethyl-2-methyleneindoline and 2-allylidene-1,3,3-trimethylindoline groups. The symmetric TAM$^+$ dyes with styryl-ring pendants are expected to possess elongated conjugation from the N$^+$ center of the FB ring to the phenyl ring. However, these LTAM molecules were not successfully obtained from the reactions of Fischer's base with the substituted cinnamaldehydes.

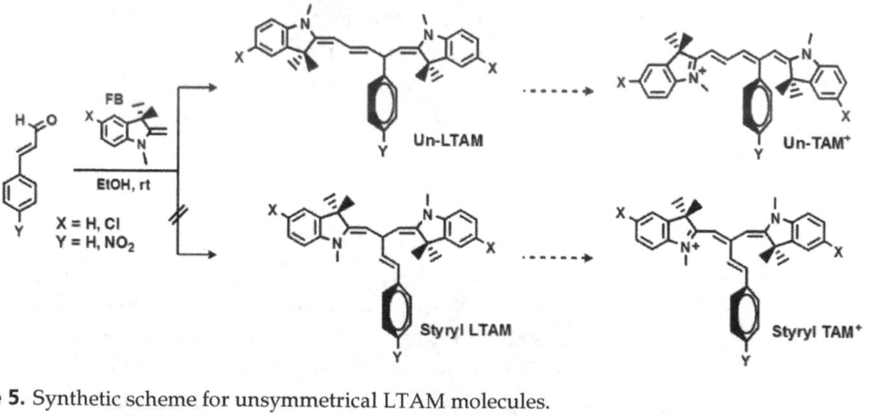

Figure 5. Synthetic scheme for unsymmetrical LTAM molecules.

Experimentally, Un-LTAM molecules were formed as the sole product and no symmetrical LTAM dyes were formed. This suggests that the Michael-type addition of the second molecule of FB occurs on the δ-carbon and not on the β-carbon of the extended α,β-unsaturated iminium salts that were formed from the reaction of FB and cinnamaldehydes. The mechanistic processes for the formation of Un-LTAM molecules are shown in Fig. 6.

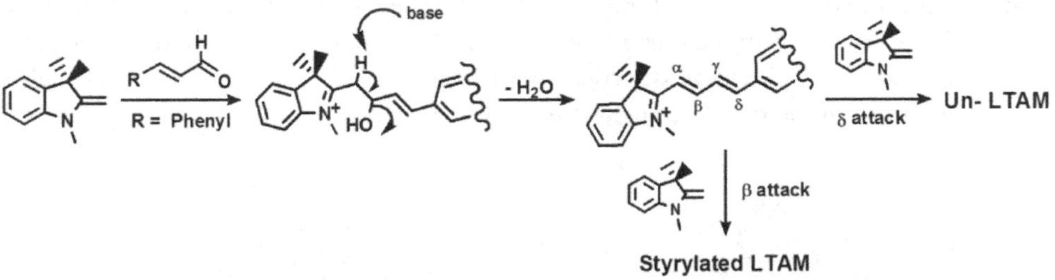

Figure 6. Mechanistic processes of the Michael-type addition of a FB molecule to the β and δ carbon of the α, β, γ, and δ-unsaturated iminium salts to form symmetrical and unsymmetrical LTAM dyes, respectively.

Melting points, yields, and other characteristic data of the prepared LTAM and Un-LTAM molecules are summarized in Table 1.

Compound	LTAM moecule		M.p. (°C)	Yield (%)	Colour
	X	Y			
LTAM 1	H	H	146	75	white
LTAM 2	H	p-Cl	168	69	pink
LTAM 3	H	p-NO₂	162(dec.)	67	reddish
LTAM 4	Cl	H	189	82	white
LTAM 5	Cl	p-NO₂	182	89	orange
LTAM 6	Cl	m-NO₂	118(dec.)	72	reddish
LTAM 7	Cl	2-Cl, 5-NO₂	204	65	orange
LTAM 8	Cl	p-(N)	218-219	89	white
LTAM 9	Cl	m-(N)	189-190	69	white
LTAM 10	Cl	p-OMe	146	57	white
LTAM 11	Cl	p-CHO	157	80	pale orange
LTAM 12	Benzo[e]	H	181-182	21	pale green
LTAM 13	Benzo[e]	m-(N)	152	48	pale lime
Un-LTAM 1	H	H	191	42	brown
Un-LTAM 2	H	p-NO₂	186	46	brown
Un-LTAM 3	Cl	H	182	53	orange
Un-LTAM 4	Cl	p-NO₂	177	55	orange

Table 1. Melting points (M.p.), yields, and colours of the prepared LTAM and Un-LTAM molecules.

3. Spectroscopic Characterization

Newly synthesized FB analogs of LTAM molecules are not expected to be fully planar because of steric crowding, rather they can be viewed as a screw or helical, and thus may possess propeller structures. Subsequently, they can adopt a conformation where all three rings are twisted in the same direction, making a right- or left-handed propeller. As an analogy to a common screw or bolt, right- and left-handed screws are nominated as *M* and *P*, respectively, as shown in the upper line of Fig. 7. The red arrows denote the direction of bond rotation, not the helical direction (blue arrows). Two rings rotate through a perpendicular conformation while one moves in the opposite direction.

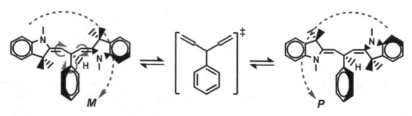

Figure 7. A new diastereomer can be formed via bond rotation (red arrows) of one diastereomer (blue arrows denote the helical direction between *M* and *P* isomers).

Theoretically, three configurational isomers, *ZE*, *EE*, and *ZZ*, can be proposed for these dyes. Since the *ZE* isomers can have *M* and *P* conformations, *ZE*-LTAM molecules can be obtained as a racemic mixture from the synthetic reaction described previously.

3.1. Diastereomeric identification for the prepared LTAM molecules by 1D ^{1}H NMR and 2D NMR experiments

The ^{1}H NMR spectra of LTAM molecules display characteristic signals (three groups) in the aliphatic region, namely a triplet and two doublets (group A) in the range of 4.20–5.40 ppm, two singlets (group B) between ~2.90 and ~3.30 ppm, and four identical singlets (group C) at 1.20–1.80 ppm. As a representative example, the ^{1}H NMR spectrum of LTAM **4** in CDCl$_3$ is shown in Fig. 8.

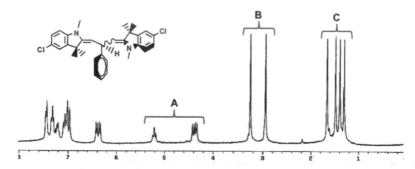

Figure 8. ^{1}H NMR spectroscopy of LTAM **4** as a representative example.

Judging from the chemical shifts of the signals in Fig. 8, group A may belong to sp^2 protons, H2a and H2a' and allylic proton H1a", and groups B and C may belong to N-Me (H10) and gem-dimethyl groups (H8 and H9), respectively. Interestingly, the gem-dimethyl groups show four well-separated singlets, indicating that these gem-dimethyl groups are diastereotopic. Thus, the gem-dimethyl groups are not identical and they have different chemical shifts in the NMR spectra. The most common instance of diastereotopic groups is when two similar groups are substituents on a carbon adjacent to a stereogenic center. These LTAM molecules may not be fully planar because of steric crowding and would thus be expected to exhibit chirality, having no stereogenic centers. No detailed discussions are not given for the resoncances in the range of 6 to 8 ppm, since the resonance in those ranges are of the general aromatic protons.

The ^{1}H and ^{13}C resonances of the LTAM molecules were assigned using COSY and one-bond ^{1}H–^{13}C correlations obtained by both direct-detection HETCOR and indirect detection HSQC experiments. COSY was used to identify peaks from the A and B rings. The HETCOR and HSQC identified the shifts of the proton-containing carbons. HMBC was used to differentiate between the two A and B rings of one half of the molecules because these rings were not identical.

The HETCOR experiment identified the carbon shifts of those carbons with protons attached through one-bond coupling between ^{1}H and ^{13}C. Correlations between the protons such as H2a, H8, H9, H10 and H1a", and their corresponding carbons are particularly useful for structure determination, as indicated in Fig. 9.

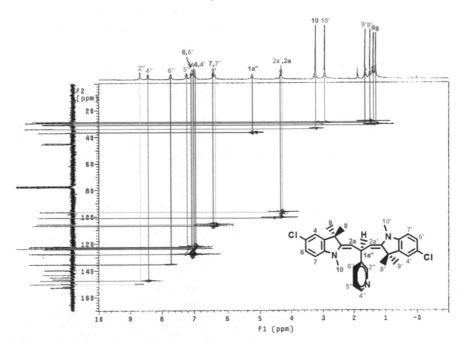

Figure 9. HETCOR of LTAM **9** in the range of of 0-170 ppm.

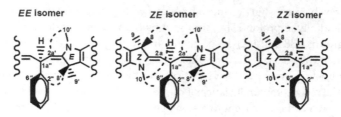

Figure 10. Structures and NOE correlations of the ZE, EE, and ZZ isomers of LTAM molecules.

The major chemical shift assignments within FB rings, A and B, of the molecules were mostly made using HMBC and NOE data. Since A and B rings are not identical, HMBC was needed to differentiate between the two groups of FB molecules. HMBC experiments have thus correlated C2 to H1a", H2a, H8, H9, and H10 of ring A and correlated C2' to H1a", H2a', H8', H9', and H10' in ring B. For the geometrical identification around the double bonds of the enamine moiety of the A and B (FB) rings, NOE experiments were carried out. Structures and NOE correlations of the ZE, EE, and ZZ isomers of the LTAM molecules are shown in Fig. 10. Determination of the configuration of the double bonds at positions 2–2a and 2'–2a' of the enamine moiety of the A and B (FB) rings was carried out by an NOE experiment. Strong NOE correlation was observed between the protons H10' (N-Me of B) at 2.97 ppm and H2a' (the same subunit) at 4.42 ppm. In addition, H10 (N-Me of A) has NOE with H2"/H6", whereas the *gem*-dimethyl, H8 and H9, exhibits NOE with H2a. These observations are compatible with a Z arrangement around the double bond of ring A.

Compound	H or C	Z Ring		E Ring		Others[a]	
		δ (H)	δ (C)	δ (H)	δ (C)	δ (H)	δ (C)
LTAM 4	2/2'	-	151.6	-	151.3	-	-
	2a/2a'	4.36(d)	97.3	4.42(d)	101.5	-	-
	3/3'	-	44.8	-	44.2	-	-
	3a/3a'	-	139.4	-	139.6	-	-
	8/8'	1.32(s)	30.6	1.50(s)	28.6	-	-
	9/9'	1.41(s)	30.8	1.68(s)	29.0	-	-
	10/10'	3.26(s)	33.5	2.95(s)	29.2	-	-
	1a"	-	-	-	-	5.22(dd)	38.7
	2"/6"	-	-	-	-	7.46(d)	127.6
LTAM 5[b]	2/2'	-	-	-	155.8		
	2a/2a'	-	-	4.42	100.5		
	3/3'	-	-	-	45.16		
	3a/3a'	-	-	-	139.6		
	8/8'	-	-	1.39	30.6		
	9/9'	-	-	1.60	28.6		
	10/10'	-	-	2.97	29.0		
	1a"	-	-	-	-	5.19(t)	38.8
	2"/6"	-	-	-	-	7.53(d)	128.8

Compound	H or C	Z Ring		E Ring		Others[a]	
		δ (H)	δ (C)	δ (H)	δ (C)	δ (H)	δ (C)
	2/2'	-	-	-	152.0	-	-
	2a/2a'	-	-	4.41(d)	100.1	-	-
	3/3'	-	-	-	44.7	-	-
	3a/3a'	-	-	-	139.6	-	-
LTAM 8[b]	8/8'	-	-	1.41(s)	29.3	-	-
	9/9'	-	-	1.51(s)	28.4	-	-
	10/10'	-	-	2.97(s)	28.6	-	-
	1a"	-	-	-	-	5.07(t)	38.3
	2"/6"	-	-	-	-	7.31(d)	122.9
	2/2'	-	151.6	-	151.2	-	-
	2a/2a'	4.30(d)	97.9	4.37(d)	102.3	-	-
	3/3'	-	44.9	-	44.3	-	-
	3a/3a'	-	139.6	-	139.9	-	-
LTAM 10	8/8'	1.30(s)	30.8	1.47(s)	29.4	-	-
	9/9'	1.39(s)	30.7	1.65(s)	28.6	-	-
	10/10'	3.25(s)	33.7	2.94(s)	28.9	-	-
	1a"	-	-	-	-	5.15(dd)	37.8
	2"/6"	-	-	-	-	7.35(d)	127.4

[a]Others denote the phenyl ring and connecting groups of the LTAM molecules.

[b]Compound that has the EE configuration as the major isomer.

Table 2. ¹H and ¹³C NMR spectral data for LTAM molecules in CDCl₃ (500 and 125 MHz, respectively).

Similarly, H2a' and H9' have NOE correlations with H10' and H2"/H6", respectively. These NOE phenomena indicate a ZE geometry around the double bonds of the enamine moieties of A and B (FB) rings, respectively. Selected ¹H and ¹³C NMR spectral data for the major diastereomer of various LTAM molecules in CDCl₃ (500 and 125 MHz, respectively) are listed in Table 2.

3.2. Thermal diastereomerization

3.2.1. Diastereomeric mixtures in equilibrium state

The ¹H NMR spectra of LTAM **4** became complicated upon thermal treatment. After approximately 2 h they exhibited three sets of signals corresponding to two other forms (a & b) in addition to the original major set, in CDCl₃, as shown in Fig. 11. Two of the double bonds in the LTAM molecules can exist as ZE, EE, and ZZ isomers, which may account for the complexity of the spectra. Their existence is most likely due to geometrical isomerism with respect to restricted rotation around the C=C double bond of the FB moiety and the C-CH(FB)(Ph) single bonds.

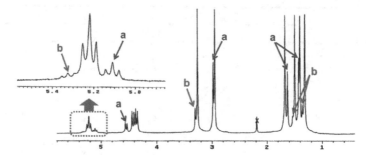

Figure 11. ¹H NMR spectra of LTAM **4** at the thermal equilibrium state, showing three sets of analog peaks (major, a and b groups).

Detailed analysis of the ¹H NMR spectra of the LTAM compounds in the thermal equilibrium state is important for determining the presence of a mixture of ZE and EE or ZZ isomers. After thermal equilibrium of LTAM **4** in CDCl₃, 2 h after sampling in an NMR tube at room temperature, three sets of complex signals were observed, namely triplet peaks at 5.0–5.4 ppm, three doublets at 4.3–4.6 ppm, four singlets at 2.8–3.4 ppm, and eight singlets at 1.3–1.7 ppm. Among these peaks, those signals assigned to the ZE isomers were a triplet at 5.22 ppm, two doublets at 4.36 and 4.42 ppm, two singlets at 2.95 and 3.26 ppm, and four singlets at 1.3–1.7 ppm, as discussed previously. The residual peaks might belong to the EE and/or ZZ isomers.

For identification of each of the diastereomers of LTAM **4** in organic solvents, 2D NMR experiments such as COSY, HMBC, and NOESY, were used at the equilibrium state. As an example, a COSY of diastereomeric mixtures of LTAM **4** in the range of 4.25–5.25 ppm in CDCl₃ is given in Fig. 12.

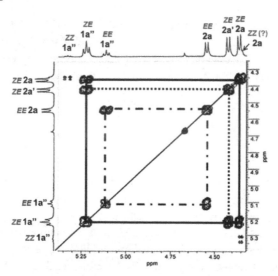

Figure 12. COSY of diastereomeric mixtures of LTAM **4** in the range of 4.25–5.25 ppm in CDCl₃.

¹H-¹H COSY in the range of 4.25–5.25 ppm showed two individual sets of H1a" and H2a/2a' protons for the diastereomeric structures ZE (solid and dot) and EE (dash-dot) isomers of LTAM 4, respectively. The methylene doublets of the ZZ isomers for this compound could not be detected due to their low concentration in the equilibrium state. Three sets of N-Me (H10 or H10') in the range of 2.8–3.4 ppm and the germinal methyl group (H8 and H9, and H8' and H9') could be easily distinguished through the visual peak ratios of the ¹H NMR.

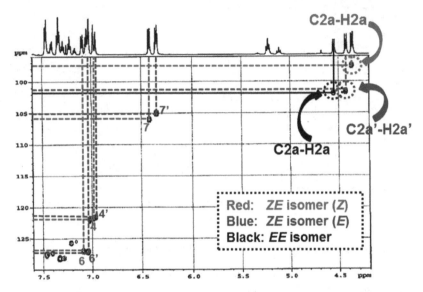

Figure 13. HSQC of diastereomeric mixtures of LTAM 4, showing one-bond correlation of C2a-H2a, C2a'-H2a' for the ZE isomer and C2a-H2a for the EE isomer in the range of 95-130ppm.

¹H–¹³C correlations were obtained by both direct-detection HETCOR and indirect detection HSQC experiments. Fig. 13 shows some of the one-bond ¹H–¹³C correlations. HSQC identifies the shifts of the carbons bearing protons of the major ZE and minor EE isomers.

More particularly, HMBC can identify which protons belong to which unit. For the major ZE isomers, HMBC experiments have correlated C2 to H1a", H2a, H8, H9, and H10 of ring A and correlated C2' to H1a', H2a', H8', H9', and H10' in ring B. The gem-dimethyls (1.50 and 1.68 ppm) are correlated with C3' at 44.2 ppm. The H2a' at 4.42 ppm is correlated to the same C3', which allow us to assign it to the same subunit (B ring). Similarly, the N-Me at 2.95 ppm, correlated to the same C3', is also in the same subunit (B ring). The other gem-dimethyl groups (1.32 and 1.41 ppm) are correlated with C3 at 44.8 ppm. The H2a at 4.36 ppm and the N-Me at 3.26 ppm are also correlated to the same C3, indicating that they are in the second subunit (A ring). The low-field HMBC of the diastereomeric mixtures of LTAM 4 is given in Fig. 14 and the high-field HMBC (< 95 ppm) are not given here.

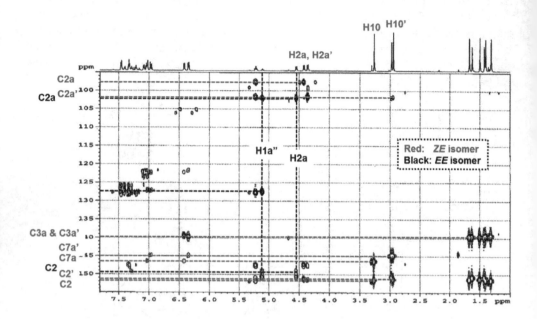

Figure 14. HMBC of the diastereomeric mixtures of LTAM **4** in the range of 95–155 ppm.

For the minor concentration *EE* isomers, the HMBC experiments correlated C2 at 149 ppm to H1a", H2a, H8, H9, and H10. Similar correlations of C3 were made for H2a, H4, H8, and H9. HMBC could further correlate C3a to H7, H8, and H9. Similar to the extremely minor *ZZ* isomers, HMBC correlated C2 to H1a", H2a, H8, H9, and H10.

Similar correlations of C3 were made to H2a, H4, H8, and H9. HMBC further correlated C3a to H7, H8, and H9. These correlations provided a clear distinction between diastereomeric isomers. H2a and *N*-Me were also coupled to the same C3, which confirms that all of the protons belong to the same isomer. As H9 or *N*-Me is coupled to C2"/C6", the protons of the aromatic ring could be identified as a substructure of each isomer.

2D NOESY showed spatial correlations for each of the diastereomeric mixtures for LTAM **4** after reaching thermal equilibrium, as in Fig. 15. Namely, the spatial correlations labeled a–f in red were detected for the *ZE* isomer, and those labeled b', d', and f' in blue are detected for the *EE* isomer. In addition, one correlation, e' in green, was detected for the *ZZ* isomer.

Unfortunately, a few of the ¹H resonance peaks for the *ZZ* isomers were able to be detected, such as an *N*-methyl singlet and, very rarely, two *gem*-dimethyl peaks. This result indicates that the LTAM molecules equilibrate in a time-dependent manner, yielding a mixture of the ZE/EE/ZZ isomers in organic solvent (CDCl₃).

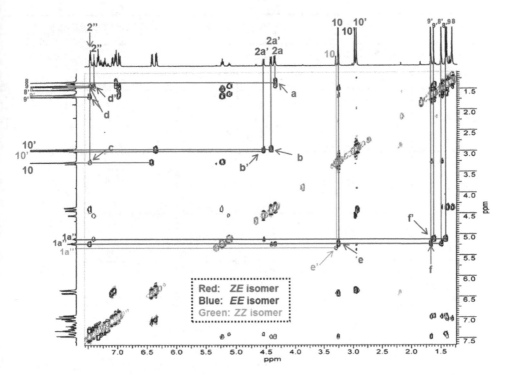

Figure 15. 2D NOESY of the diastereomeric mixtures of LTAM **4** after reaching thermal equilibrium in CDCl₃.

This NOE phenomenon indicates a ZE geometry around the double bonds of the enamine moieties of A and B (FB) rings of the major isomer. However, the minor isomer contains two symmetric Fischer's base units and 2D NOESY only showed the correlations of H2a'-H10' (blue mark, b'), H8'-H2"/H6" (blue mark, d') but no correlation of H10'-H2"/H6". This suggests that the ZE geometry around the double bonds of the enamine moieties does not exist for the minor isomer. Although the spatial correlations of H2a-H9, H10-H2"/H6" and H1a"-H10 were expected for the extremely minor ZZ isomer, the proton peaks H2a and H2a' of the ZZ isomer were too small to be correlated with other protons in the 2D NOESY experiment. One spatial correlation of H1a"-H10 (green mark, e') was detected. This NOE phenomenon indicates that three diastereomers, such as ZE, EE, and ZZ, are in equilibrium with various ratios among the diastereomeric isomers, depending on the NMR solvent used.

The NMR data for the Z or E ring of the ZE isomer suggest that the signals for the geminal dimethyl group for the EE isomer should be shifted downfield compared to those of the ZZ isomer, whereas the signals of the N-methyl groups should be shifted upfield. The geminal dimethyl group signals for the EE isomer were shifted upfield compared to those of the ZE isomer, whereas the signals of the N-methyl and methylene groups were shifted downfield.

Compound	Ring	proton	Diastereomer (ppm)		
			ZE	EE	ZZ
LTAM 1	Z	8-Me	1.31	-	1.37
		9-Me	1.41	-	1.57
		N-Me	3.28	-	3.32
		H2a	4.33	-	N/A
	E	8'-Me	1.51	1.44	-
		9'-Me	1.68	1.64	-
		N'-Me	2.97	3.00	-
		H2a'	4.41	4.53	-
LTAM 2	Z	8-Me	1.33	-	1.38
		9-Me	1.42	-	1.55
		N-Me	3.29	-	3.33
		H2a	4.31	-	N/A
	E	8'-Me	1.52	1.46	-
		9'-Me	1.68	1.64	-
		N'-Me	2.99	3.01	-
		H2a'	4.36	4.47	-
LTAM 4	Z	8-Me	1.32	-	N/A
		9-Me	1.41	-	N/A
		N-Me	3.26	-	N/A
		H2a	4.36	-	N/A
	E	8'-Me	1.50	1.43	-
		9'-Me	1.68	1.63	-
		N'-Me	2.95	2.98	-
		H2a'	4.42	4.55	-
LTAM 5	Z	8-Me	1.30	-	N/A
		9-Me	1.39	-	N/A
		N-Me	3.22	-	3.28
		H2a	4.31	-	N/A
	E	8'-Me	1.41	1.39	-
		9'-Me	1.65	1.60	-
		N'-Me	2.94	2.97	-
		H2a'	4.32	4.42	-
LTAM 10	Z	8-Me	1.30	-	N/A
		9-Me	1.39	-	N/A
		N-Me	3.25	-	3.28
		H2a	4.30	-	N/A
	E	8'-Me	1.47	1.41	-
		9'-Me	1.65	1.60	-
		N'-Me	2.94	2.96	-
		H2a'	4.37	4.49	-
LTAM 12	Z	8-Me	1.67	-	N/A
		9-Me	1.77	-	N/A
		N-Me	3.44	-	N/A
		H2a	4.47	-	N/A
	E	8'-Me	1.90	1.83	-
		9'-Me	2.06	2.04	-
		N'-Me	3.08	3.11	-
		H2a'	4.52	4.60	-

Table 3. Selected ^{1}H resonances for some of the diastereomeric LTAMs.

Deshielding of the *gem*-dimethyl proton of the *EE* and the N-methyl proton of the *ZZ* isomer may be due to their relative proximity to the benzene ring, as indicated in the *ZE* isomers. Selected ¹H resonances for the diastereomeric LTAMs are listed in Table 3.

3.2.2. Dynamic behavior of LTAM molecules

Interestingly, the stability of these molecules in solution depends upon the solvent media. Namely, they are inert in polar organic solvents such as acetone and DMSO, but they are unstable in nonpolar solvents such as benzene, THF, and chloroform. They equilibrate time-dependently into a mixture of *ZE/EE* or *ZE/EE/ZZ* isomers, depending on the solvents used, as shown in Fig. 16.

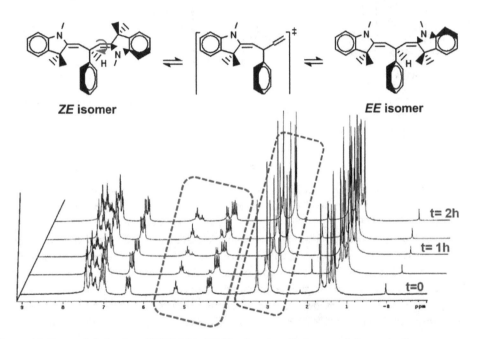

Figure 16. Dynamic behavior of LTAM **1** in CDCl₃, showing diastereomeric isomerization.

It has been reported (Keum et.al. 2008) that FB-analogs of LTAM molecules have very characteristic ¹H NMR resonance patterns in the range of 1.0–5.4 ppm as a result of three consecutive protons (H2a, H2a', and H1a''), two N-methyl, and four diastereotopic *gem*-dimethyl (8- and 9-Me) groups. Therefore, these characteristic peaks can be used to discriminate each of the diastereomers. (Ma et al., 2012) For example, ¹H NMR data of the 3-pyridinyl LTAM **9** showed the expected features (A → B) of resonances, *viz.* one triplet at 5.25 ppm, two doublets at 4.31 and 4.34 ppm, two singlets at 2.95 and 3.25 ppm, and four singlets at 1.30–1.65 ppm. In contrast, the spectra of 4-pyridinyl LTAM **8** showed very interesting features in the range of 2.90–5.40 ppm. This compound showed one triplet at 5.07 ppm, a doublet at 4.41 ppm, and a singlet at 2.97 ppm, as shown in C of Fig. 17.

The isomerization pattern of LTAM **8** is quite surprising because no FB-analog of MG showed a LTAM **8**-like feature (C → B) in the range of 2.90–5.40 ppm. Based upon quantum mechanical calculations (Keum et al., 2010), *ZE* would be expected to predominate over *ZZ* and *EE* in all media. Experimentally, the *ZE* isomers of LTAM compounds generally predominant in all organic media examined. The relative energy differences between the minor *EE* and extremely minor *ZZ* were 0.08 and 0.26 kcal/mol in CDCl₃ and DMSO-d₆, respectively. In addition, both of the spectra, A and C, converged to that displayed by B approximately 2–3 h after mixing the LTAM molecules with CDCl₃ in the NMR tube.

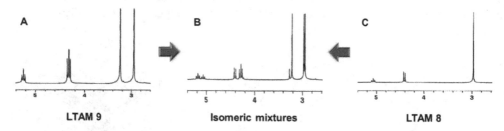

Figure 17. Characteristic proton resonance of the LTAM molecules in the range of 2.80~5.40 ppm in CDCl₃ (A: *ZE*, B: mixture after reaching thermal equilibrium, and C: *EE* diastereomer); arrows show the directions of isomerization.

3.2.3. Determination of diastereomeric ratios at equilibrium state

Since the characteristic peaks can be used to discriminate each of the diastereomers, the equilibrium ratios among these diastereomeric isomers in various organic solvents can be determined by ¹H NMR spectra. This is based on the intensities of either the *N*-methyl or *gem*-dimethyl signals corresponding to the three diastereomeric isomers at the equilibrium state. In some cases, the intensity of the H1a" proton of the central carbon can be used. Since the *N*-methyl peaks show a well-separated singlet, it is more convenient to measure the ratio of the isomers. The cause of the isomerization of the LTAM compounds at room temperature is unclear. They belong to the group of conjugated enamine compounds. Enamine-imine tautomerism (C=C-NH and CH-C=N) may regulate *ZE* isomerization.

For most of the LTAM molecules examined that are listed in Table 1, the *ZE* isomers are the most stable and they at the equilibrium state for almost 100% of the time in polar solvents (E_T(30) > 42) and 60–80% in non-polar solvents (E_T(30) < 42) at room temperature. The minor *EE* minor and extremely minor *ZZ* isomers at the equilibrium state were 18–44% and 0–11% in nonpolar solvents, respectively, depending on the molecules examined. The percent ratios among the diastereomeric isomers of LTAM molecules in the thermal equilibrium states vary according to the molecules examined and solvents used.

However, some LTAM molecules, such as LTAMs **3**, **5**, and **8**, are exceptional. Surprisingly, the pure *EE* isomers are obtained, unlike for the LTAM molecules described previously. These exceptional compounds contain a resonance-electron withdrawing (-R) substituent, particularly on the para-position of the phenyl ring. This indicates that substituents such as

p-NO$_2$, p-CHO, or p-(N) on the phenyl ring make the *EE* isomer more stable than the *ZE* isomer, which is predicted to be more stable theoretically. These are summarized in Table 4.

Further detailed studies are needed to determine how the isomerization occurs and what causes the unusual stability of a certain diastereomer.

Compound	Solvent	Percent ratio (%)			K_{eq}[a]	Note[b]
		ZE	*EE*	*ZZ*		
LTAM 4	CDCl$_3$	60.2	28.3	11.4	1.52	
	Aceton-d_6	68.5	25.9	5.60	2.17	*ZE*
	DMSO-d_6	91.5	8.5	-	10.76	
LTAM 5	CDCl$_3$	61.0	34.1	4.90	0.52	
	Aceton-d_6	59.9	34.8	5.30	0.53	*EE*
	DMSO-d_6	33.2	68.8	-	2.07	
LTAM 8	CDCl$_3$	63.3	31.3	5.40	0.46	
	Aceton-d_6	61.3	32.8	5.90	0.49	*EE*
	DMSO-d_6	36.4	63.6	-	1.75	
LTAM 9	CDCl$_3$	64.0	31.1	4.80	1.78	
	Aceton-d_6	62.4	32.0	5.60	1.66	*ZE*
	DMSO-d_6	59.2	40.8	-	1.45	
LTAM 10	CDCl$_3$	67.1	27.4	5.50	2.03	
	Aceton-d_6	65.8	27.7	6.50	1.92	*ZE*
	DMSO-d_6	80.7	13.0	6.30	4.18	

[a]K_{eq} is the ratio of [*ZE*]/[*EE*+*ZZ*] or [*EE*]/[*ZE*+*ZZ*], depending on the identity of the major diastereomer.

[b]The major isomer in the solid state.

Table 4. Percent ratios among the diastereomeric isomers of LTAM molecules at thermal equilibrium states.

3.2.4. Free energy change of activation, $\Delta G^{\ddagger}$

The rate constants for the formation of the *EE* (and *ZZ)* isomers were measured from a plot of ln(A−A^0) versus time (in min), according to the peak-intensity of the central proton at ~5.15 ppm. As an example, excellent linearity was obtained with r = 0.999 and n = 6. The k_{obs} and half-life ($t_{1/2}$) for the isomerization of LTAM 4 were 5.95 × 10^{-4} s^{-1} and 19.4 min, respectively. The first-order rate constant is a sum of the rate constants for the backward and reverse reactions. From the rate constant obtained at room temperature, the obtained one-temperature $\Delta G^{\ddagger}$ value (Dougherty et al., 2006) for the *ZE* → *EE* isomerization of LTAM 4 in CDCl$_3$ was found to be 21.8 kcal/mol. Similarly, the rate constants for the diastereomeric isomerization of LTAM molecules were measured and the one-temperature $\Delta G^{\ddagger}$ values of all of them were obtained, using the equation 1 (Dougherty et.al. 2006) given below:

$$\Delta G^{\ddagger} = 4.576 \, [10.319 + \log (T/k)] \text{ kcal/mol} \tag{1}$$

These are summarized in Table 5.

Compound	k_{obs} x10^{-4} s^{-1}	$t_{1/2}$ (min)	Linearity		$\Delta G^{\ddagger}_{ZE \to EE}$	$\Delta G^{\ddagger}_{EE \to ZE}$
			r	n		
LTAM 4	5.95	19.4	0.999	6	21.8	-
LTAM 5	0.43	297	0.999	3	-	23.4
LTAM 8	2.72	42.5	0.999	6	-	22.3
LTAM 9	4.66	24.8	0.994	4	22.0	-
LTAM 10	2.48	46.5	0.998	5	22.0	-
LTAM 11	4.72	24.5	0.998	5	-	22.4
LTAM 13	17.9	6.47	0.998	3	21.2	-

Table 5. Rate constants and $\Delta G^{\ddagger}$ values for the ZE/EE isomerization of LTAM molecules in CDCl$_3$.

It has been previously reported that the ZE isomerization of imines and their tautomeric isomers, enamines, has a very high energy barrier ($\Delta G^{\ddagger}$ = 23 kcal/mol), unless the process is strongly accelerated by either acid/base catalysts or by push-pull substituents. (Liao & Collum, 2003). The isomerization rate was found to be slow on the NMR time-scale.

3.3. UV-Vis spectroscopy of various forms of LTAM molecules

FB analogs of TAM$^+$ dyes were obtained from the reaction of FB analogs of LTAM molecules with DDQ in the presence of HCl, followed by separation of the deep blue form from the product mixtures by column chromatography in MC/MeOH (7:1). A reaction of TAM$^+$ with an inorganic base such as NaOH gives the carbinol form of the LTAM molecule. Only the TAM$^+$ cation shows deep coloration, in contrast to the LTAM and carbinol derivatives. This difference arises because only the cationic form has extended π-delocalization, which allows the molecule to absorb visible light.

The colored forms, TAM$^+$, of the prepared LTAM and Un-LTAM molecules have absorption maxima at 580–705 and 350–420 nm in ethanol for the x- and y-band, respectively. The carbinol form was detected at 325–385 nm in basic media. The leuco form of these molecules decomposed in HCl-saturated EtOH to form conjugated molecules observed at 385–435 nm. UV-Vis spectral data in CDCl$_3$ of the colored and decomposed forms LTAM **4**, and Un-LTAM **4**, as representative examples, are shown in Fig. 18.

UV-Vis spectral data for various forms of LTAM and Un-LTAM molecules, compared to those of commercial TAM$^+$ dyes, are summarized in Table 6.

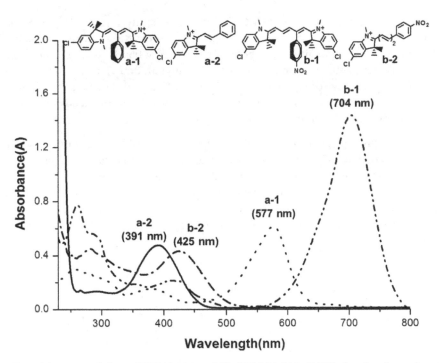

Figure 18. UV-Vis spectral data of LTAM 4 (a) and Un-LTAM 4 (b) in EtOH, showing the various forms such as the TAM⁺ (a-1 and b-1) and decomposed forms (a-2 and b-2).

| Compound[a] | Various structural forms (λ_{max}) | | | | |
| | Leuco- (a) | Carbinol- (b)[b] | TAM⁺- (c)[c] | | Decomposed dye (d) |
			x-band	y-band	
Crystal Violet[d]	265	266	585	-	-
Malachite Green[e]	265	265	620	430	-
LTAM 1	284	343	609	426	385
LTAM 4	296	327	578	370	391
LTAM 5	298	-	588	380	390
LTAM 8	302	363	595	377	318
LTAM 9	295	-	556	368	382
LTAM 11	298	-	591	-	403
Un-LTAM 1	322	381	693	412	420
Un-LTAM 2	326	383	686	414	415
Un-LTAM 3	324	368	686	353	435
Un-LTAM 4	326	384	704	417	425

[a]Names of compounds are the same as in Table 1. [b]The carbinol denoted a hydroxylated TAM⁺ dye.

[c]Symbols (x-, y-band) are adopted from Ref. (Ernest et al., 1989) [d,e]Data for acetonitrile.

Table 6. UV-Vis spectral data for various forms of LTAM and Un-LTAM compounds.

In the UV-Vis spectral data of Table 6, MG and crystal violet dyes show absorption maxima at 620 and 430 nm for the x- and y-band, respectively, whereas the absorption maxima of the vinyl-log of MG are red-shifted for both the x- and y-bands, i.e., 651 and 488 nm, respectively. This suggests that the vinyl effects of a vinyl unit may, to a large extent, behave like extended conjugation for both the x- and y-bands. Chemical skeletons for the N~N⁺ and C(phenyl)~N⁺ responsible for the x- and y-band, respectively, in the absorption spectra of various TAM⁺ dyes are shown in Fig. 19.

Structurally, the FB analogs of symmetric and unsymmetric TAM⁺ dyes in this work can be characterized as Cy3 and Cy5 dyes, respectively, as closed-chain cyanines.(Ernst et al., 1989) It was reported that Cy3 is maximally excited at 550 nm and maximally emits at 570 nm in the orange-red part of the spectrum, whereas Cy5 is maximally excited at 649 nm and maximally emits at 670 nm, which is in the red part of the spectrum. Therefore, the x-band of the Un-TAM⁺ are expected to be higher than 650 nm and 550 nm, for the y- and x-band, respectively.

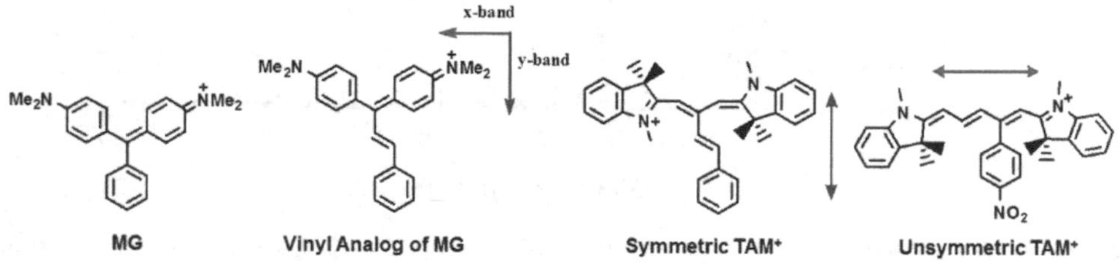

Figure 19. Chemical skeleton for the N~N⁺ and C(phenyl)~N⁺ responsible for the x- and y-band, respectively, in the absorption spectra of various TAM⁺ dyes.

From the reaction of Un-LTAM **4** with $HClO_4$, the decomposed product {5-chloro-1,3,3-trimethyl-2-((1E,3E)-4-(4-nitrophenyl)buta-1,3-dienyl)indolium perchlorate} was isolated, brown, yield 57%, M.p.= 257–258 °C, IR (KBr) 3072, 2984, 2934, 1707, 1596, 1340, and 1086 cm^{-1}, ¹H NMR (DMSO-d_6) δ 1.76 (6H, s), 4.03 (3H, s), 7.37 (1H, d, J = 15.3 Hz), 7.66 (1H, dd, J = 10.2, 15.3 Hz), 7.73 (1H, d, J = 9.0 Hz), 7.79 (1H, d, J = 15.3 Hz), 7.92 (2H, d, J = 6.9 Hz), 7.95 (1H, d, J = 9.0 Hz), 8.09 (1H, s), 8.33(1H, dd, (J = 10.2, 15.3 Hz), and 8.33(2H, d, J = 6.9 Hz).

4. Solid state structure

4.1. LTAM molecules

The X-ray crystal structure of LTAM **12,** as a representative example, displays an orthorhombic crystal system with space group $Pna2_1$, with a residual factor of $R_1 = 0.0517$. ORTEP diagrams of LTAM **12,** showing atom numbering, are provided in Fig. 20(a).

For LTAM **12,** the C7-C8 and C7-C24 distances are 1.512 and 1.513 Å, respectively, i.e., typical lengths for C-C single bonds, and the enamine C8-C9 and C24-C25 bonds are 1.335 and 1.335 Å, respectively, which are typical lengths for C=C bonds. The LTAM **12** molecules possess three-bladed propeller conformations, similar to earlier reports for various nonhetaryl LTAM dyes. The inter-plane angles between the aromatic rings A-B, A-C, and B-C in LTAM **12** are 81.29, 87.79 and 86.02°, respectively, as shown in Fig. 20(b).

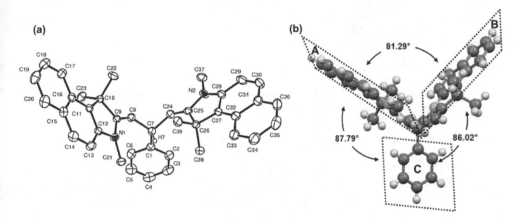

Figure 20. ORTEP diagrams with atom numbering scheme (a) and the propeller shape (b) of LTAM **12**, showing the inter-plane angles.

Selected bond lengths and bond angles for the LTAM molecules are listed in Table 7.

Ring	Bond length (Å) and Bond angle (°)	LTAM molecule			
		4	5	9	12
A	C7-C8	1.513	1.511	1.515	1.512
	C8-C9	1.330	1.341	1.334	1.335
	C9-N1	1.410	1.398	1.400	1.405
	C8-C9-N1	129.42	122.8	129.72	127.47
B	C7-C24	1.507	1.509	1.507	1.513
	C24-C25	1.332	1.334	1.330	1.335
	C25-N2	1.409	1.416	1.407	1.396
	C24-C25-N2	123.06	123.0	123.20	122.94
Others	C7-C8-C9	130.99	128.3	131.06	129.21
	C7-C8-H8	114.5	115.8	114.5	115.4
	C8-C7-H7	107.6	108.6	107.7	112.10
	C7-C24-C25	127.42	129.7	127.68	128.50
	C7-C24-H24	116.3	115.2	116.2	115.8
	C24-C7-H7	107.6	108.6	107.7	107.2

Table 7. Selected bond lengths and bond angles of LTAM molecules.

The dihedral angles H8-C8-C7-H7 and H24-C24-C7-H7 in LTAM **12** are 172.06° (θ_1) and 176.41° (θ_2), respectively. The inter-plane angles and dihedral angles for the LTAM molecules are given in Table 8.

Compound	Interplane angles[a] (º)			Dihedral angles[b] (º)	
	ring A-B	ring B-C	ring A-C	θ_1	θ_2
LTAM 1	79.8	74.9	84.8	164.1	149.8
LTAM 4	60.1	75.7	83.2	178.9	158.9
LTAM 7	77.9	85.6	80.2	152.9	139.8
LTAM 8	135.4	72.8	72.8	163.2	163.2
LTAM 9	60.0	77.3	83.5	156.1	179.0
LTAM 11	44.83	74.32	72.92	174.62	151.68
LTAM 12	81.29	86.02	87.79	172.06	176.41

[a]Symbols (A-B) and numbering systems are as indicated in Fig. 20(b).
[b]Symbols (θ_1 and θ_2) are the dihedral angles H(8)-C(8)-C(7)-H(7) and H(24)-C(24)-C(7)-H(7), respectively.

Table 8. Inter-plane angles and dihedral angles for LTAM molecules in the solid state.

The C(7)=C(8) double bonds of the two 5-chloro Fischer's base moieties have *EE* configurations in **1**. In contrast, in **2**, the C(14)=C(15) and C(2)=C(3) double bonds of the two 5-chloro Fischer's base moieties belong to the *ZE* configuration. The *EE* (for **1**) and *ZE* (for **2**) isomers formed as the sole product in each case, despite the fact that three isomers, namely *ZE*, *EE*, and *ZZ*, are possible for these dyes which result from the reaction of excess 5-chloro Fischer's base and 4- and 3-pyridine carboxaldehyde.

Compound **12** is stacked so that a dimer is formed in the unit cell of the crystal. The packing in the unit cell of LTAM **12** is distinct, as can be seen in Fig. 21.

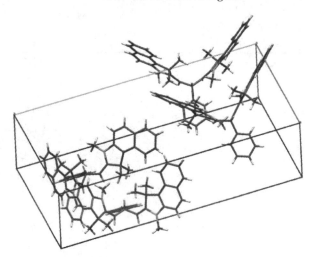

Figure 21. Molecular packing of LTAM **12**, showing the formation of a dimer.

4.2. Un-LTAM molecules

The Un-LTAM **4** was only successfully crystallized from acetone. Unfortunately, crystal growth was unsuccessful for the remainder of the Un-LTAM molecules. Selected bond lengths and bond angles are listed in Table 9.

Ring	Bond length (Å)		Bond angle (°)	
	N1-C2	1.397	C2-N1-C7a	111.7
A	N1-C10	1.439	C2-N1-C10	123.9
	C2-C7"	1.345	C2-C7"-H7"	117.1
	N2-C2'	1.408	C2'-N2-C7a'	111.3
B	N2-C10'	1.445	C7a'-N2-C10'	124.5
	C2'-C11"	1.324	C2'-N2-C10'	123.1
	C1"-C10"	1.516	C9"-C10"-C11"	110.92
	C9"-C10"	1.516	C1"-C10"-C11"	110.01
Conneting group	C10"-C11"	1.509	C9"-C10"-C11"	110.92
	C8"-C9"	1.328	C8"-C9"-C10	127.29
	C7"-C8"	1.443	C11"-C2'-N2	122.42

Table 9. Selected bond lengths and bond angles of Un-LTAM **4**.

The X-ray crystal structure of Un-LTAM **4** shows a triclinic crystal system with space group *P-1*. An ORTEP diagram of Un-LTAM **4**, including the atom-numbering scheme, is shown in Fig. 22(a).

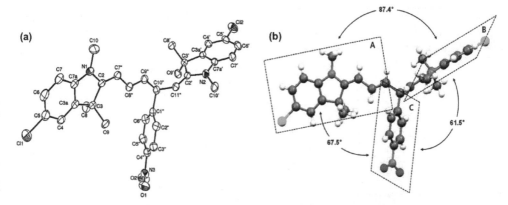

Figure 22. ORTEP diagrams with atom numbering scheme (a) and the propeller shape (b) of Un-LTAM **4**, showing the inter-plane angles.

The C9"-C10" and C10"-C11"'' distances are 1.516 and 1.509 Å, respectively, typical for C–C single bonds. The C7"-C8" single bond distance, however, was 1.443 Å, which is shorter than a typical single bond and longer than a typical double bond. This is perhaps due to conjugation since the length (1.47 Å) of the central single bond of 1,3-butadiene is approximately 6 ppm shorter than that of the analogous single bond (1.53 Å) in butane. The two enamine C2=C7" and C2'=C11", and C8"=C9" double bonds were 1.354, 1.324, and 1.328 Å, respectively, which are typical C=C bond lengths. In the crystal, the three aromatic rings of **4** are linked to three different layers, *viz.* a vinyl FB, a FB, and a phenyl group. The unsymmetrical molecule is a distorted version of the well-known three-bladed propeller conformation. (Keum et al., 2011). The inter-plane angles of the aromatic rings A–B, A–C, and B–C in Un-LTAM **4** are 87.4°, 67.5°, and 61.5°, respectively (Fig. 22(b)).

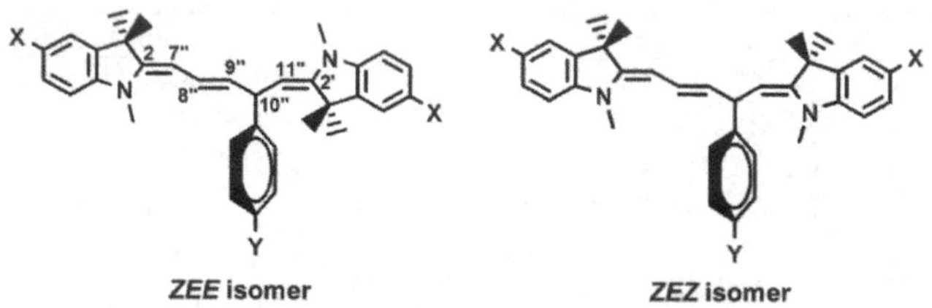

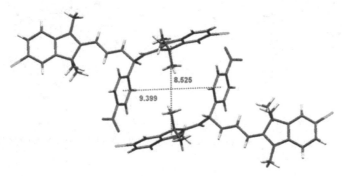

ZEE isomer   **ZEZ isomer**

Figure 23. Chemical structures of Un-LTAM , with ZEE and ZEZ configurations.

The double bonds C2=C7", C8"=C9", and C2'=C11" of Un-LTAM **4** have EEE configurations. The EEE isomers of these LTAM dyes are formed as the sole product in all cases, even though there are three possible isomers, the two other diastereomers being the ZEE and ZEZ isomers, as shown in Fig. 23. Generally, the central carbon-carbon double bond of these LTAM dyes is expected to have an E configuration.

Although the presence of the ZEE and ZEZ diastereomers was generally expected to be found in organic solvents, none of these isomers were detected, unlike for the LTAM molecules examined previously.

Fig. 24 shows the molecular packing diagram of Un-LTAM **4**, showing the formation of the dimer, which is stacked in an alternating fashion in the unit cell of the crystal. The intermolecular distances in the dimer are 8.53 and 9.40 Å, for the FB and phenyl rings, respectively.

Figure 24. Molecular packing diagram of Un-LTAM **4**, showing formation of the dimer.

5. Conclusion

Novel Fischer's base analogs of LTAM and Un-LTAM molecules and their corresponding TAM⁺ dyes have been successfully developed. ^1H and ^{13}C NMR assignments for the prepared LTAM molecules have been completed by 1D and 2D NMR experiments, including DEPT, COSY, HSQC, HMBC, and NOESY. The geometry of the double bond was ZE in most cases, as measured directly by NOESY. The EE and ZZ isomers have C_2 symmetry, and hence, the two FB rings of these isomers are identical. Therefore, the ^1H NMR spectra of the EE and ZZ isomers are expected to be relatively simple compared to those of the ZE isomer. The novel LTAM molecules exist as a single isomer (ZE or EE) in the solid phase and they are equilibrated with other isomers in organic solvents. The percent ratios among the diastereomeric isomers of LTAM derivatives in the thermal equilibrium states vary according to the molecules examined and solvents used.

UV-Vis spectral data shows various structural forms of the LTAM and Un-LTAM molecules, such as (a) leuco-, (b) colored TAM⁺, (c) carbinol-, and (d) decomposed-forms, similar to the commercially known TAM⁺ dyes, such as MG, crystal violet, etc. Particularly, UV-Vis spectroscopic data for the Un-TAM⁺ dyes showed absorptions in the near-IR region.

X-ray crystal analysis showed that the ZE isomers were generally formed with a so-called three-bladed propeller conformation. These isomers stacked to form a dimer or double dimer. However, the EE isomers were also formed specifically for the LTAMs **3**, **5**, **8**, and **11**, which have a resonance-electron withdrawing (-R) group at the para-position of the phenyl ring. Further analysis of a variety of substituted LTAM molecules is required to determine what makes the diastereomer structures change in the solid state.

Author details

Sam-Rok Keum, So-Young Ma and Se-Jung Roh
Korea University at Sejong Campus, South Korea

Acknowledgement

This research was supported by Basic Science Research Program through the National Research Foundation of Korea (NRF) funded by the Ministry of Education, Science and Technology (No.2012003244) and partly by the Brain Korea 21 project.

6. References

Anslyn, E.V. & Dougherty, D.A. (2006). *Modern Physical Organic Chemistry*, University Science Books, U.S. pp. 365-367.

Balko L, Allison J. The direct detection and identification of staining dyes from security inks in the presence of other colorants, on currency and fabrics, by laser desorption mass spectrometry. *J. Foren. Sci.* 2003, *48*: 1172–8.

Bartholome, D. & Klemm, E. (2006). Novel Polyarylene–Triarylmethane Dye Copolymers, *Macromolecules*, Vol. 39 pp. 5646-5651.

Cho BP, Yang T, Lonnie R, Blankenship L, Moody JD, Churchwell M, Beland FA. Culp S. *J. Chem. Res. Toxicol.* 2003, *19*, 285.

Ernst, L. A.; Gupta, R. K,; Mujumdar, R. B. & Waggoner, A. S. (1989). Cyanine dye labeling reagents for sulfhydryl groups, *Cytometry* Vol. 10, pp. 3-10.

Fengling, S. F. & Xiaojun, P. X. (2005). Heptamethine Cyanine Dyes with a Large Stokes Shift and Strong Fluorescence: A Paradigm Excited-State Intramolecular Charge Transfer. 2. *J. Am. Chem. Soc.* Vol. 127, pp. 4170-4171.

Gessner, T. & Mayer, U. (2005). Triarylmethane and Diarylmethane Dyes, *Ullmann's Encyclopedia of Industrial Chemistry*, Weinheim: Wiley-VCH, doi:10.1002/14356007.a27-179

Indig GL, Anderson GS, Nichols MG, Bartlett JA, Mellon WS, Sieber F. Effect of molecular structure on the performance of triarylmethane dyes as therapeutic agents for photochemical purging of autologous bone marrow grafts from residual tumor cells. *J. Pharm. Sci.* 2000, *89*: 88–99.

Kraus, G. A.; Jeon, I.; Nilsen-Hamilton, M.; Awad, A. M.; Banerjee J. & Parvin, B. (2008). Fluorinated Analogs of Malachite Green: Synthesis and Toxicity, *Molecules,* Vol. 13, No. 4, pp. 986-994; doi:10.3390/molecules13040986

Keum, S. R.; Roh, S. J,; Lee, M. H.; Saurial, F. & Buncel, E. (2008). ^{1}H and ^{13}C NMR assignments for new heterocyclic TAM leuco dyes, (2Z,2'E)-2,2'-(2-phenyl propane-1,3-diylidene)bis(1,3,3-trimethylindoline) derivatives. Part II. *Magn. Reson. Chem.* Vol. 46, pp. 872–877.

Keum, S. R.; Roh, S. J.; Kim, Y.N.; Im, D.H. & Ma, S. Y. (2009). X-ray crystal structure of hetaryl leuco-TAM dyes, (2Z,2'E)-2,2'-(2-phenylpropane-1,3-diylidene) bis(1,3,3-trimethyl indoline) derivatives. *Bull. Korean Chem. Soc.* Vol. 30, pp. 2608–2612.

Keum, S. R.; Roh, S. J,; Ma, S. Y.; Kim, D. K. & Cho, A. E. (2010). Diastereomeric isomerization of hetaryl leuco-TAM dyes, (2Z, 2'E)-2,2'-(2-phenyl propane-1,3-diylidene) bis(1,3,3-trimethylindoline) derivatives in various organic solvents. *Tetrahedron,* Vol. 66, pp. 8101–8107.

Keum, S. R.; Lee, M. H.; Ma, S. Y.; Kim, D. K. & Roh, S. J. (2011). Novel unsymmetrical leuco-TAM, (2E, 2'E)-2,2'-{(E)-4-phenylpent-2-ene-1,5-diylidene}bis(1,3,3-trimethyl indoline) derivatives: synthesis and structural elucidation. *Dyes and Pigments,* Vol. 90, pp. 233–238.

Liao, S. & Collum, D. B. (2003). Lithium Diisopropylamide-Mediated Lithiations of Imines: Insights into Highly Structure-Dependent Rates and Selectivities, *J. Am. Chem. Soc.* Vol. 125, pp. 15114-15127.

Li, Z.; Duan, Z.; Kang, J.; Wang, H.; Yu, L. & Wu, Y. (2008). A simple access to triarylmethane derivatives from aromatic aldehydes and electron-rich arenes catalyzed by FeCl₃, *Tetrahedron* Vol. 64, pp. 1924-1930.

Ma, S. Y.; Kim, D. K.; Lim, H. Y.; Roh, S. J. & Keum, S. R. (2012). Unusual Stability of Diastereomers of the Isomeric Pyridine-based Leuco-TAM Dyes 2,2'-(2-(Pyridin-4 or 3-yl) propane-1,3-diylidene)bis(5-chloro-1,3,3-trimethylindoline), *Bull. Korean Chem. Soc.* Vol. 33, pp. 681-684.

Nair, V.; Thomas, S.; Mathew, S. C. & Abhilash, K. G. (2006). Recent advances in the chemistry of triaryl- and triheteroarylmethanes. *Tetrahedron,* Vol. 62, pp. 6731-6747.

Özer, I. & Çaglar, A. (2002). Protein-mediated nonphotochemical bleaching of malachite green in aqueous solution. *Dyes & Pigments,* Vol. 54, pp. 11-16.

Plakas, S. M.; Doerge, D. R.; Turnipseed, S. B. In Xenobiotics in Fish; Kluwer Academic and Plenum Publisher: NY. 1999; pp. 149.

Keum, S. R.; Ma, S. Y.; Kim, D. K.; Lim, H. W. & Roh, S. J. (2012). Novel dimeric leuco-TAM dyes,   1,4-bis{(1E,3Z)-1,3-bis(1,3,3-trimethylindolin-2-ylidene)propan-2-yl}benzene derivatives: Structure and spectroscopic charactierization. *Journal of Molecular structure,* Vol. 1014, pp. 25-32.

Keum, S. R.; Ma, S. Y.; Kim, D. K.; Lim, H. W. & Roh, S. J. (2012). Unsymmetric leuco-TAM dyes, (2E, 2'E)-2,2'-{(E)-4-phenylpent-2-ene-1,5-diylidene}bis(1,3,3-trimethylindoline) derivatives. Part II: X-ray crystal structure[*]. *Dyes and Pigments,* Vol. 94, pp. 490–495.

Chapter 5

Phenotiazinium Dyes as Photosensitizers (PS) in Photodynamic Therapy (PDT): Spectroscopic Properties and Photochemical Mechanisms

Leonardo M. Moreira, Juliana P. Lyon, Ana Paula Romani, Divinomar Severino, Maira Regina Rodrigues and Hueder P. M. de Oliveira

Additional information is available at the end of the chapter

1. Introduction

Oscar Raab demonstrated, in 1900, that the light incidence on dyes can induce cell death [1]. A photosensitizer is a chemical compound that is activated by light of a specific wavelength that leads to tumor destruction [2]. Indeed, Photodynamic Therapy (PDT) is considered to have its origin in 1900 with the classical experiments by the german scientist Oscar Raab. Raab noticed that the exposure of *Paramecium caudatum* to acridine orange and later subjection to light resulted in death of this organism. Raab and his supervisor Hermann von Tappeiner later coined the term "photodynamic therapy" and applied PDT successfully for the treatment of cutaneous tumors using eosin. From that concept, photodynamic therapy (PDT) [3,4,5,6], as we known today, was founded. Since then, the development of other studies, culminating with those performed by Dougherty and co-workers resulted in a non-invasive technique for cancer treatment and other diseases [7,8]. In fact, precancerous cells, certain types of cancer cells and microbial infections can be treated this way.

Interesting data regarding the application of PDT against several diseases have been reported, since the employment of this therapy in different diseases has increased significantly. In fact, PDT has been used with phenotiazinium [methylene blue (MB) and toluidine blue] as photosensitizers against AIDS-related Kaposi's sarcoma, promoting complete sarcoma remission with excellent cosmetic results [9]. PDT with MB (and LED as light source), which is a very inexpensive system, has been applied against *Leishmania*, promoting significant reduction in the size of the lesions, diminishing the parasitic load in the draining lymph node and healing the lesions in hamsters experimentally infected with L. *amazonensis* [10]. This therapeutic alternative is very interesting due to the resistance of this organism to pentavalent antimonials (SbV), which constitutes the mainstay pharmacological alternative for leishmaniasis, due to emergence of drug resistance [11].

Tumor, which is also called neoplasm or blastoma, is the abnormal growth of tissues. Sick cells with genetic disturb develop more rapidly than the normal cells, which provokes the development of the tumor (that can be malign or non-malign cells). When the growth of the tumor is a very fast and chaotic process, with tendency to arrive in other organs, generally is a malign tumor [4]. Cancer is the general name of all malign tumors. This term cancer is originated from latin and means "crab". This name is due to the tendency of the tumor to be fixed in several biological tissues, which is correlated to the ability of the crab to be fixed in various surfaces [4].

Interestingly, the PDT procedure is easily performed in a physician's office or outpatient setting, which favors the application of this therapy in several environments, since PDT does not need great structural pre-requisites. In this context, it is important to notice that multicenter randomized controlled studies have demonstrated high efficacy and superior cosmetic outcome over standard therapies [12]. In fact, several cosmetic methodologies have been developed with PDT, such as resurfacing. For many non-oncologic dermatological diseases, such as *acne vulgaris*, viral warts and localized scleroderma, case reports and small series have confirmed the potential of PDT [12]. After the development of topical photosensitizers 5-aminolevulinic acid (ALA) or its methyl ester (MAL), PDT has gained worldwide popularity in dermatology, since these drugs do not induce prolonged phototoxicity as the systemic photosensitizing hematoporphyrin derivatives do [12]. PDT has essentially three steps. First, a light-sensitizing liquid, cream, or intravenous drug (photosensitizer) is applied or administered. Second, there is an incubation period of minutes to days. Finally, the target tissue is then exposed to a specific wavelength of light that activates the photosensitizing medication.

More than one million cases of skin cancer were diagnosed during 2008 in the U.S.A. and its worldwide incidence has risen throughout the last four decades. Squamous cell carcinoma (SCC) is the second most frequent skin cancer, only after basal cell carcinoma (BCC) [13]. In the 20th century, SCC was mainly linked to occupational sun exposure, whereas in the last decades the strongest link has been to ultraviolet (UV) radiation. On one hand, UVB exposure leads to direct DNA damage by pyrimidine dimer formation. On the other hand, UVA induces formation of reactive oxygen species which indirectly also cause DNA damage. Other factors such as the phototype, the genetic predisposition or the immune response are also involved in the carcinogenic process [13].

It is also important to notice that photoantimicrobial agents, that is, chemical compounds that exhibit increased inactivation of microorganisms when exposed to light, have been known also for over a century [14]. While there are several studies regarding the use of photosensitizers against bacterial and viral targets, the clinical use of photosensitizers in antimicrobial therapy has been developed very slowly through small scale trials. This is particularly a surprise considering the efficacy exhibited, especially by cationic photosensitizers, against pathogenic drug-resistant bacteria such as methicillin-resistant *Staphylococcus aureus* and vancomycinresistant *Enterococcus faecium* [14]. Furthermore, the exponentially increasing threat of microbial multidrug resistance has highlighted antimicrobial photodynamic inactivation (APDI) as a promising alternative treatment for localized infections [15]. APDI involves the direct application of the PS to the infected tissue rather than being injected intravenously, as the usual procedure for cancer treatment with PDT [15].

The photodynamic process involves photophysical and photochemical steps, which can be applied with several aims, such as therapies against cancer or infections. PDT light sources include laser, intense pulsed light, light-emitting diodes (LEDs), blue light, red light, and many other visible lights (including natural sunlight). Photosensitizer drugs may become activated by one or several types of light. The optimal light depends on the ideal wavelength for the particular drug used and target tissue.

Electron and energy transfer in the excited state govern the efficiency of a variety of photoinduced processes, including photosynthesis, light to energy conversion in semiconductor devices, cell damage induced by solar exposition and photodynamic action [16,17,18]. It is well reported that photophysical behavior of a dissolved dye depends on the nature of its environment, i.e., the solvent influences the spectra characteristics of the solute molecules [19]. Several factors influence the visible spectral behavior of dissolved dye molecules, especially the solvent's polarity and its hydrogen-bond donor/acceptor capacities [19]. The properties can be determined by the solvent dielectric constant, ε, and solvatochromic parameters. The strong solvatochromic behavior can be observed for dye molecules with large dipole moment changes during transitions between two electronic states. The solvent can differentially stabilize the ground and/or the excited state in polar and non-polar solvents [19].

The series of phenothiazine [thionine, methylene blue (MB), azure A (AZA) and azure B (AZB)] derivatives (Fig. 1) are positive dyes used as a model for phototherapeutic agent as well as for dye sensitized solar energy converter [20,21] due to their appropriate biological, chemical, photochemical and photophysical properties [22,23,24].

The intersystem cross quantum yield and the singlet oxygen formation for MB is 0.52 [25,26,27,28,29], the triplet lifetime is higher, approximately 3.0 μs, in air saturated aqueous solution, and up to 50 μs in nitrogen saturated aqueous solution. The singlet excited state has a lower lifetime, approximately 1,400 ps (Table 1), and it is due to the higher internal conversion and triplet formation, with a fluorescence quantum yield of 0.04 in methanol [30,31,32,33]. In addition, MB and and MB derivatives that have been used as photosensitizers in PDT showed a good biocompatibility (appropriate citotoxicity and phototoxicity) when used *in vitro* to attack key organelles in cells [14,21].

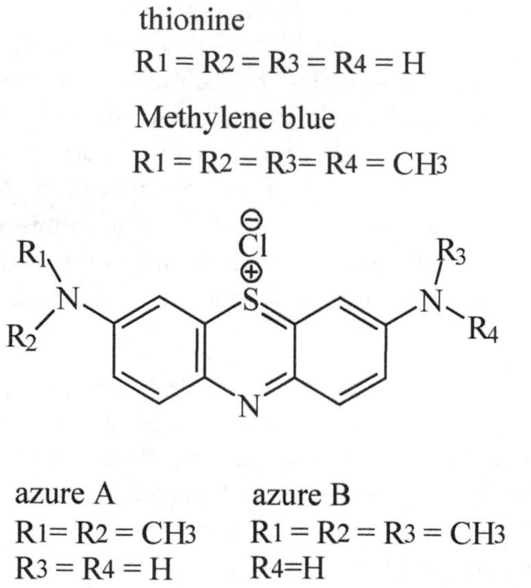

thionine

R1 = R2 = R3 = R4 = H

Methylene blue

R1 = R2 = R3= R4 = CH3

azure A   azure B

R1= R2 = CH3   R1 = R2 = R3 = CH3

R3 = R4 = H   R4=H

Figure 1. Thionine derivatives.

medium: water	
Dye	τ (ps)
Thionine	314,40
Nile blue	372,75
Azure A	421,28
Azure B	1268,56 (48,89%) e 306,00 (51,11%)
Toluidine blue	2179,65 (66,72%) e 358,03 (33,28%)
Methylene blue	328,84
medium: ethanol	
Dye	τ (ps)
Thionine	848,84
Nile blue	1170,03
Azure A	776,43
Azure B	724,78
Toluidine blue	643,89
Methylene blue	465,96

Table 1. Values of lifetime (τ) of some dyes at 25°C.

The Fluorescence decays of dyes were obtained by single-photo-counting technique. The excitation source was a Tsunami 3950 Spectra Physics titanium-sapphire laser, pumped by a Millenia X Spectra Physics solid state laser. The laser was tuned that a third harmonic generator BBO crystal (GWN-23PL Spectra Physics) gave the 292 nm excitation pulses that were directed to an Edinburgh FL900 spectrometer. The spectrometer was set in L-format configuration, the emission wavelength was selected by a monochromator (680 nm), and emitted photons were detected by a refrigerated Hamamatsu R3809U microchannel plate photomultiplier. The software provided by Edinburgh Instruments was used to analyze the individual decays. The quality of the fit was judged by the analysis of the statistical parameters reduced-χ^2 and Durbin-Watson, and by the inspection of the residuals distribution.

The dyes stock solutions were prepared in ethanol (6.0×10^{-5} M) and aliquots of these stock solutions were added, via a calibrated Hamilton microsyringe, to volumetric flasks containing water or ethanol, and the solutions were stirred for 30 minutes. The final concentrations of dyes were 1.0×10^{-6} M. All measurements were performed at 25°C using a cuvette with 0.2 cm of optical path.

The excited state lifetime depends on the solvent [34,35]. The dependence of the lifetime on the viscosity and solvent dielectric constant indicates that the dye excited state deactivation process is slow as the medium viscosity increases. This effect is related to the partial inhibition or the higher friction on the dye substitute groups rotation, such as $-CH_3$, $-NH_2$, $-N(CH_3)_2$ and $-N(CH_2CH_3)_2$ [36]. The lifetime values are in agreement with the results reported in the literature. Lee and Mills [37] showed the lifetime values for methylene blue aqueous solutions (358 ± 20 ps). Grofcsik et al [34] measured the lifetime of Nile blue excited state and oxazine 720 in different solvents at 20 °C. The thionine dye photophysics is well known [38]. In an aqueous solution, thionine has a fluorescence lifetime of 320 ± 60 ps when excited at 610 nm [37,39]. In organic solvent, the increase of the thionine fluorescence lifetime (450 ps in ethanol and 760 ps in terc-butilic alcochol) results in a increasing of the fluorescence quantum yield [38]. The thionine lifetime differences observed in an aqueous medium and ethanol is quite high, which shows the effect of microenvironment polarity on the excited state decaying [38]. In our experiments, in an aqueous medium thionine has a useful lifetime of 314.4012 ps, which is in agreement with the results presented in the literature.

The Nile blue lifetime in ethanol and water are 1420 and 418 ps, respectively. These results are higher than those that we found in our work. However, it should be taken into account that the temperatures used in our experiments are different from those whose results are different. It was shown that the lifetime of Nile blue depends on the temperature due to the intermolecular charge transfer [34,35,36]. This charge transfer process is facilitated by the presence of NH2 groups in molecule structure, such as on the Nile blue structure, which may change the lifetime values. Grofcsik et al [34] studied these probes in different solvents where it was observed that there is a relationship between the solvent permissivity and the excited state lifetime. It was shown that the lifetime is higher in nonpolar solvents, where protic solvents decreases the excited state lifetime. This behavior was observed in both dye molecules that were studied, which have a similar chemical structure.

Grofcsik et al [34,35] have shown that there is a relationship between the excited state lifetime of Nile blue and Oxazine 720 with the acidity of the medium. As the hydrogen ion concentration increases it is observed a decrease of the excited state lifetime [40]. It was also observed for methylene blue, azure A, azure B and azure C [41]. The reason for the rapid decay in acid medium is due to the formation of dications from the monocations reaction in the excited state with hydrogen ions. These results indicate that the reaction in the excited state the additional protons are located on the nitrogen atom of the ring and not on the terminal amine groups [40]. It is believed that for other compounds the results may be the same due to the similarity in the chemical structure of such molecules.

Dutt et al [42] studied the fluorescence lifetime of cresila violet, Nile blue, oxazine 720 and Nile red, using different solvents, such as alcohols, polyalcohols, amides and some aprotic solvents. The authors showed that the lifetime values for these dyes are approximately 3.5 ns for n-alcohols, which are higher than that for the Nile blue (1.62 ns in ethanol). This result is in agreement with our studies. When it is considered the behavior of bipolar solute in polar solvents, the hydrodynamic and dielectric contribution must be taken into account [42]. However, it is not well known how to measure these hydrodynamic and dielectric contributions individually. In the case of the four dyes, when in the presence of amides and aprotic solvents, as described above, the contributions are reasonably described by the hydrodynamic friction, where to describe the rotating relaxation in the presence of n-alcohols; the dielectric friction must be included.

Chen et al [43] studied the quantum yield of the methylene blue singlet oxygen as a function of the medium pH values. The authors showed that the protonated acid (3MBH2+) triplet state is similar to the base (3MB+) triplet state, and the quantum yield of the singlet oxygen formed is much higher in basic medium than that in acidic medium. The singlet oxygen formation increases as the pH of the medium is increased, while the singlet state lifetime decay the triplet state formation do not depend on the pH changes. It can be explained by the population decay rate of the singlet state due to the internal conversion to the fundamental state, and the intersystem crossing to the triplet state, which are much higher that the protonation rate [43,44]. Also [43] studied the behavior of methylene blue, 1,9-dimethyl-methylene blue and toluidine blue in aqueous medium and methanol. The triplet state formation and the singlet oxygen quantum yield in water were very similar to that for methylene blue and for 1,9-dimethyl-methylene blue. The kinetic studies results for the singlet state decay of methylene blue in water and in methanol were 0.37 and 0.62 ns, respectively, where for toluidine blue the results were 0.28 and 0.40 ns, respectively. In the case of methylene blue the decay useful life of the singlet excited state in methanol is approximately two times higher than in water. The authors showed that there is no influence of the solution concentration on the singlet state lifetime, where the differences on the lifetime decays that were observed in water and methanol are not related to the methylene blue dimerization in water. The methylene blue lifetime decay decreases with the increase of the dielectric constant of protic solvents due to the interaction of the methylene blue with the polar solvent [45]. In protic alcohols and in aqueous solutions the methylene blue excited state lifetime is higher than of the fundamental state. Therefore, the differences between the singlet and triplet states decrease as the relaxation rate is increased. In the presence of aprotic solvents, such as acetone, acetonitrile, and dimethyl sulfoxide, the dipole excited state is lower in the fundamental state, where the energy differences observed is higher and the relaxation lifetime is longer [46].

The use of these dyes as singlet oxygen photosensitizer in PDT, as well as tumor cells removal are being investigated [47,48,49,50]. It is known that under laser irradiation in the presence of photosensitizer dyes, the tumor cells undergo necrosis or apoptosis and the rate of tumor cell removal through apoptosis increases [51,52,53]. This behavior has been related to the presence of singlet oxygen in the tumor cells [54,55]. The increase of cell removal through apoptosis is of great importance in the PDT treatment [50,56]. There are no side effects in the cell removal through apoptosis because it is a controlled cell removal process, where there is no inflammation of the laser irradiated tissue. In some cases changes in the PDT mechanism has been observed, type I via free radical and type II via oxygen singlet, which could be related to the interaction among the dyes and the cellular system [57,58,59,60]. These changes involve the aggregation of two or more dye molecules in the same site [61].

2. Photodynamic Therapy (PDT): Mechanism of action

Selective tumor destruction without damaging surrounding healthy tissues can be reached by using PDT, which is treatment, activated by light, which requires the combination of three elements: a photosensitizer, visible or near-infrared light, and oxygen [62,63,64,65,66,67,68,69]. However, the precise mechanisms of PDT are not yet fully understood but two general mechanisms of photoinduced damage in biomolecules have been proposed: Type I and Type II [62,70,71]. Type I is the photodynamic mechanism in which the excited molecule induces radical formation that causes damage to biological targets (membranes, proteins and DNA), and an electron transfer event is the initial step [16]. In Type I mechanism, the photosensitizer in the excited state interacts directly with a neighbor molecules, preferentially O_2, producing radicals or radical ions through reactions of hydrogen or electron transfer [72]. Frequently, these radicals react immediately with the O_2 generating a complex mixture of reactive oxygen species (ROS), such as hydrogen peroxide, superoxide radical and hydroxyl radical, which are capable to promote oxidation a great number of biomolecules [62].

It is believed that $^1O_2$ produced through type II reaction is primarily responsible for cell death. It is known that several factors including the PS, the subcellular localization, the substrate and the presence of O_2 contribute to this process [71]. The lifetime of $^1O_2$ is very short (approximately 10-320 nanoseconds), limiting its diffusion to only approximately 10 nm to 55 nm in cells [73]. Type I photoreaction of some PSs are primarily responsible for sensitization through radical formation under hypoxic conditions. In the presence of oxygen, $^1O_2$ mediates photosensitization process, but the supplemental role of H_2O_2, OH• and O_2• must also to be considered. Only substrates situated very close to the places of ROS generation will be firstly affected by the photodynamic treatment because the half-life of $^1O_2$ in biological systems is under 0.04 µs and its action radius being lower than 0.02 µm [71]. This assumption is due to the fact that ROS are highly reactive and present a very short half-life. Type II is the photodynamic mechanism in which the photooxidation is mediated by singlet oxygen ($^1O_2$), where an energy transfer reaction from the photoexcited molecule to molecular oxygen is the initial step [16,62]. The process involves the excitation of the photosensitizer from a ground singlet state to an excited singlet state, where intersystem crossing to a longer-lived excited triplet state will occur. It is also important to point out that molecular oxygen is present in tissue with a ground triplet state. When the photosensitizer and an oxygen molecule are in proximity, an energy transfer can take place that allows the photosensitizer to relax to its ground singlet state, and create an excited singlet state oxygen molecule. Additionally, energy is transferred from triplet protoporphyrin IX to triplet oxygen, resulting in singlet ground state protoporphyrin IX and excited singlet oxygen, which reacts with biomolecules, which can damage some cells in the treatment area. Singlet oxygen is the usual name associated to the three possible excited electronic states immediately superior to the ground state of molecular oxygen (triplet oxygen) [3].

Due to the short half-life and diffusion distance of singlet oxygen in aqueous media, PDT can be considered a highly selective form of cancer treatment, as only the irradiated areas are affected, provided that the photosensitizer is nontoxic in the absence of light [74]. This combination of light/photosensitizer/oxygen as a mode of disease treatment has expanded from an initial focus on cancer tumors to include application in certain non-neoplastic diseases including age-related macular degeneration (AMD), coronary heart disease, periodontal diseases, and microbial infections [75].

Singlet oxygen is a very aggressive chemical species and will very rapidly react with any nearby biomolecules, being that the specific targets depend directly on the physical-chemistry properties of the photosensitizer used in the photodynamic process, which will result in no desired side effects, such as destructive reactions that will kill cells through apoptosis or necrosis. Therefore, depending on whether Type-I or Type-II mechanisms take place, the therapeutic efficiency of PDT may be completely altered. Therefore, the ratio of apoptotic versus necrotic cell death in tumors treated with PDT may depend on the competition between electron and energy transfer in the reaction site [16].

Oxidative stress generated by the photodynamic action occurs because in biological systems the singlet oxygen presents significantly low lifetimes, where the lifetimes of the singlet oxygen is lower than 0.04 µs, implying that its radius of action is also reduced, being usually lower than 0.02 µm [3]. Reactive oxygen species (e.g. hydroxyl radicals or superoxide) are their high reactivity and low specificity with a broad spectrum of organic substrates [76]. Various methods have been employed for the generations of hydroxyl radicals such as O_3/UV, H_2O_2/UV, TiO_2 photo-catalysis and photo assisted Fe(III)/H_2O_2 reaction.

3. Photosensitizers

3.1. Phenothiazinium dyes

The phenothiazinium dyes were first synthesized in the late 19th century—e.g. both Methylene Blue (Caro) and Thionin (Lauth) in 1876—during what might be considered to be a "gold rush" period of chemical experimentation after the discovery of the first aniline dyes [77]. Among photobactericidal compounds, the phenothiazinium photosensitizers methylene blue (MB) and toluidine blue O (TBO) have often been used as lead structures, being effective photosensitizers with singlet oxygen quantum yields of approximately 0.40

Phenotiazinium Dyes as Photosensitizers (PS) in Photodynamic Therapy (PDT): Spectroscopic Properties and Photochemical Mechanisms

121

and exhibiting low toxicity levels in mammalian cells [14]. Members of the phenothiazine class are known to cross the blood-brain barrier and to be relatively nontoxic [78,79].

The biomedical use of phenothiazinium dyes has begun with specimen staining for microscopy by various medical scientists, among whom were famous scientists such as Romanovsky, Koch and Ehrlich. The idea of structure—activity relationships in stains developed in this era, particularly by Paul Ehrlich, laid the foundations for modern medicinal chemistry, and these principles should be followed by those attempting the properly organized photosensitizer synthesis [77]. Cellular uptake is determined by a combination of charge type/distribution and lipophilicity, both of which characteristics may be controlled by informed synthesis. Due to the expansion of PDT into the antimicrobial milieu, a far greater scope for photosensitizer design exists now. For example, in the field of blood product disinfection, an ideal candidate photosensitizer would be effective in the inactivation of bacteria, viruses, yeasts and protozoan, but would remain non-toxic and non-mutagenic in a human recipient. It is hardly surprising that none of the currently available agents fits all of these criteria [77].

Phenothiazinium dyes are cationic compounds with high redox potential that interacts with visible light inactivating several kinds of pathogenic agents in fresh plasma. Phenothiazinium dyes present great reactivity with the proteins and lipoproteins (cell membranes) and nucleic acids. These cationic compounds have limited capability to permeate the cell membrane as function of their elevated hydrophilic character [80]. Phenothiazinium dyes present significant action against encapsulated virus and some virus without capsule, such as parvovirus B19. As function of its genotoxic action, the employment of phenothiazinium dyes is prohibited in several countries, such as Germany[80]. On the other hand, the Methylene Blue is a highly hydrophobic compound with higher chemical affinity to the nucleic acids, which denotes its potential to application against virus.

Phenothiazinium dyes are photocytotoxic, and can cause photoinduced mutagenic effects [81]. In living systems, DNA acts as an important target for phenothiazinium dyes. It has been proved that these dyes can photosensitize biological damage. Azure B (AZB) is an easy available phenothiazinium dye, and has been widely employed both in metal determination and DNA staining detection. Owing positive charges on its molecular structure, AZB can bind to the DNA polyanion in living systems through electronic interactions. So, the study of the interaction of AZB with DNA in vitro is of importance.

Methylene Blue, MB (Figure 2) is a phenothiazinic dye current applied in PDT as therapeutic agent or photosensitizing compound. MB has a recognized antimicrobial effect in the dark (citotoxicity property) which can be increased, at oxygenated environment, by the incidence of light with a wavelength corresponding to its electronic absorption band [82,83].

Methylene Blue is a well-known photochemical oxidant. The photoreduction reaction of this dye by various types of electron donors has been studied quite often, and in most cases an electron transfer mechanism was proposed for explaining the observed results [84].

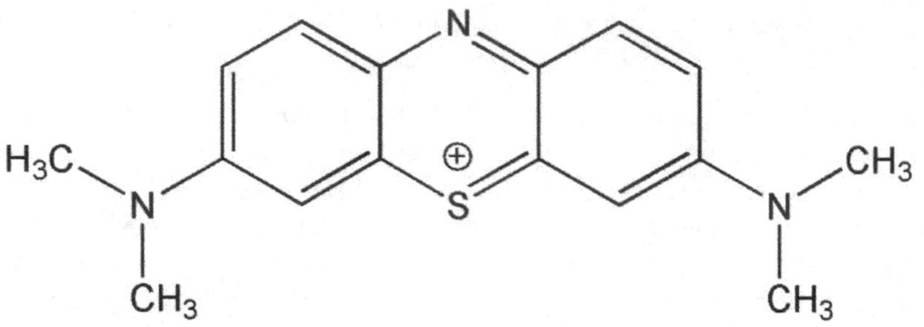

Figure 2. Structure of Methylene Blue

This molecule is particularly interesting for application in PDT due to its known physical chemical properties. For example, MB is a positive charged dye with three aromatic rings (6-members) very soluble in ethanol. It is already used clinically in humans for the treatment of metahemoglobinemia, without significant side effects. Besides these characteristics, MB presents a quantum yield of singlet oxygen formation around 0.5, with a low reduction potential, intense light absorption in the region of 664 nm in water (within the phototherapeutic window). Also, it displays a high photodynamic efficiency causing apoptosis of cancer cells, by mono or polychromatic light excitation). Currently, MB is used by several european agencies for disinfection of blood plasma, due to its efficiency in photodynamic inactivation of microorganisms such as viruses [85], including HIV, hepatitis B and C [86,87].

MB has been clinically used as a photosensitizer drug for PDT in the treatment of different types of tumors [88]. Phototherapeutic application examples include treatment of bladder cancer, inoperable esophagus tumors, skin virulence, psoriasis and adenocarcinomas [89]. Additionally, an important point to be considered is the extremely low cost of a treatment based on this dye compared with other available photo-drugs.

Although MB possesses a positive charge and the planar structure with delocalized charge, it has a tendency to form dimers, trimers or type H aggregated systems in the presence of certain additives, cell organelles or solvents, for example, water [90,91,92,93]. The development of self-aggregates compromises its photodynamic activity, impairing the production of singlet oxygen, principal phototoxic species in PDT. In self-aggregated states auto-quenching processes occur where the excited monomers have the energy suppressed by collisions with other monomers that constitute the aggregate [94,95,96,97,98].

Often, treatment protocols require unusual preparation methods, or conditions that may have many distinct characteristics of the most ideal conditions. One example is that the MB in diluted aqueous solution, with concentration around 2×10^{-5} mol L^{-1}, is found in monomeric form. However, its uses in topical treatments require concentrations higher than 6×10^{-2} mol L^{-1}, where self-aggregation and its consequences are significant [82].

Therefore, it is important to investigate the phenomena of MB self-aggregation present in solvent mixtures and / or interaction with biomolecules [90]. This study aims to investigate changes in MB spectroscopic properties caused by self-aggregate formation induced by solvent mixtures.

The MB is an oxazinic dye soluble in water or alcohol. It presents a quantum yield of oxygen singlet formation of about 0.5 and low reduction potential [25]. It is a dye with low toxicity, which absorb in the UV-visible light ($\lambda_{máx}$ = 664 nm; solvent: water) and shows good photodynamic efficiency to kill cancer cells, which can be excited by monochromatic and polychromatic light within the therapeutic window [82]. It is a hydrophobic dye, which forms aggregates when in the presence of aggregation agents such as polyelectrolyte, or when in the presence of solvents that induces the aggregation process. The aggregate formation changes photosensitization efficiency, decreasing the amount of singlet oxygen produced by light stimulation. The most important application of methylene blue (MB) is its use in PDT as a photosensitizer agent, in oncology and potentially in the treatment of other diseases, such as Leishmaniosis.

Teichert et al.[99] used *Candida albicans* strains that are resistant to the conventional treatment of *Candida* infections, which were collected from HIV-positive patients. These strains were inoculated in the oral cavity of rats that, subsequently, were submitted to the topic application of 1 mL of Methylene Blue at concentrations of 250, 275, 300, 350, 400, 450 and 500 µg mL^{-1}. After 10 minutes of dye application, the authors employed the diode laser with wavelength of 664 nm with potency of 400 mW (687.5 seconds), resulting in an energetic density of 275 J/cm^2 [100]. After one unique application, it was realized microbial culture exam of the respective samples and the individuals were sacrificed to the histological analysis of the tongue. The results obtained in this procedure demonstrated a complete elimination of the microorganisms, when the dye concentrations of 450 e 500 µg mL^{-1}were employed. In the histological analysis, the rats that were treated with PDT had no inflammatory signals. The tongues of the control group rats presented high level of infection by *Candida* which was located in the keratin layers [100]. The respective authors concluded that the PDT is a potential alternative to the treatment of the fungi infection, emphasizing, as advantages of this technique, its topic character, simple methodology and, mainly, the unspecific characteristic of PDT, i.e., the possibility of to be applied to a great number of microorganisms. Moreover, PDT can be applied several times without risk of selection of resistant yeasts [100].

Azures A, B and C, are examples of photosensitizer agents, which have the cationic derivatives, such as the Azure Bf4. The organic ions can interfere through fluorescent radiation absorption that is emitted by excited molecules, resulting in a photobactericidal effect on the *Staphylococcus aureus* and *Enterococcus faecium* colonies. This behavior is related to the light stimulation wavelength because the organic compounds present in the system absorb electromagnetic radiation. However, only organic compounds that present double bond conjugated system, such as azure A, B or C, are capable to absorb the visible light radiation.

It was observed that red visible light (600-700 nm) and nearinfrared are the wavelengths that can penetrate the human skin. The phenothiazinic dyes, such as Azures, absorb light in such wavelengths with high intensity. They show the formation of aggregates due to the presence of aggregation agents such as polyelectrolytes, or due to the presence of solvents that favors the aggregate formation, such as water. The aggregate formation changes the photosensitization process efficiency, decreasing the amount of singlet oxygen produced by the light stimulus. The self-aggregation phenomenon can be minimized by adding charged groups in the dye structure, which results in an electrostatic repulsion interaction, increasing the hydrophilic behavior of the dye, such as Azure B and Azure BF₄.

Azures are phenotiazine compounds. This class of dye has low toxicity in the dark, constant composition, being synthesized with high yield. Azures present great selectivity to the tumor cells and significant photo stability, being not maintained in the body for long interval of time. These dyes can be applied through endovenous and topic ways. Azures present high bactericide ability, being very auspicious compounds to be applied as photosensitizes in PDT, especially due to their favorable photodynamic properties and low cost [101,102]. Azure dyes (including Azure B) are recalcitrant compounds used in the textile industry. For instance, Azure B has been used in a selective assay for detecting lignin peroxidase, the oxidative enzyme with the highest redox potential produced by white-rot fungi [103,104].

Azure B is a very sensitive dye and extremely susceptible to detect slight alterations in its chemical environment, presenting significant solvatochromic processes. Physico-chemical properties of Azure B have motivated the employment of Azure B as a chromogenic reagent for the spectrophotometric determination of several compounds, which are relevant to biological and environmental chemistry such as periodate [105]. This cited method is simple and rapid, offering advantages of sensitivity and wide range of determinations, without involvement of any stringent reaction conditions, being successfully applied to the determination of periodate in solution and in several river water samples. In its time, Azure-C (AZC), and related phenothiazine compounds has been widely used for accelerating the oxidation of NADH, but not in connection to the NAD+ reduction process.

Thionine has been a subject of many studies, as for example in a photochemical and electrochemical biosensor [106,107,108,109,110] and in photovoltaic cells [111]. Thionine is a positively charged tricyclic heteroaromatic molecule, which has been investigated for its photoinduced mutagenic actions [112,113], toxic effects, damage on binding to DNA [114] and photoinduced inactivation of viruses [115]. Thionins consist of 45–47 residues bound by three to four disulfide bonds, which includes α1-purothionin, βPTH, and β-hordothionin (βHTH) [116,117,118].

It has the ability to immobilize proteins and DNA and act as molecular adhesive [119]. Biophysical and calorimetric studies with three natural DNAs of varying base compositions, have shown the intercalative binding and high affinity of thionine to GC rich DNAs [120]. Thionine presented a high preference to the alternating GC sequences followed by the homo GC sequences contained in different synthetic polynucleotides [121]. AT polynucleotides presented a lower binding affinities but the alternating AT sequences had higher affinity compared to the homo stretches. The intercalation and the sequence of specific intercalative binding of thionine were shown by fluorescence, viscosity experiments and circular dichroic studies, respectively [121].

Studies based on absorbance, fluorescence, circular dichroic spectroscopy, viscosity, thermal melting and calorimetric techniques were used to understand the binding of thionine, with deoxyribonucleic acids of varying base composition, where strong binding of thionine to the DNAs were shown. Strong hypochromic and bathochromic effects and quenching of fluorescence were observed that showed strong binding of thionine to the DNAs [97]. The binding process is exothermic, which is associated to a large positive entropy changes and a negative enthalpy, and it showed that nonelectrostatic contributions are very important for the association of thionine to DNA. Studies on the interaction of thionine with sodium dodecylsulfate (SDS) micelles have shown that thionine binding affinity to SDS micelles was decreased with increasing temperature due to the thermal agitation [122].

The spectroscopic characteristics of thionine aggregates have shown that it depends on the concentration of thionine and on the chemical nature of the solvent [123]. Two peaks can be observed, at 597 nm and at the lower wavelength side of the 597 nm peak, and they related to the monomeric species and to the aggregate formation, respectively [124]. The understanding of the thionine aggregation process is very important for some application, such as in photovoltaic cells, where the reverse homogeneous redox reaction can be inhibited due to the presence of a surfactant in the system. The presence of a surfactant interferes in the thionine aggregation and polymerization process [125].

Several works about sensors have shown that the changing of spectroscopic and electrochemical properties of organic molecules, such as Toluidine Blue O (TBO), a phenothiazine dye, may indicate that there are some interaction with mediators and biological molecules [126,127,128]. Photochemical and electrochemical properties of TBO have been used to develop new photovoltaic devices for energy conversion and storage [129]. The aggregation behavior of such dyes in phase solution can be studied by using several optical rotation and circular dichroism techniques, as it can be seen in some studies of the interaction of TBO molecules on the DNA surface [130]. It was suggested that both intercalative and electrostatic interactions of TBO with DNA, where it was pointed out that the electrostatic interaction play an important role on the formation of the bridged structure of TBO with DNA [131].

TB can also be used as an oxygen radical inactivation, biological sensitizer and complexing agent in biological systems avoiding pathological changes [132]. Due to its low toxicity and high water solubility in salt form, which has an intense absorption peak in the visible region [133], it has been used in pharmaceutical formulations for cancer treatment [134]. Studies on the micellar solutions have shown that the aggregation properties and distribution behavior of toluidine blue in the presence of surfactant depend on the electrostatic interaction. In the case of surfactin, a natural surfactant, TB molecules can be located in the palisade of surfactin micelle [135].

Nile blue (NB) belongs to a class of molecules whose basic framework is that of a benzophenoxazine, a class which also includes Nile red, a phenoxazinone, here termed red

Nile blue (RNB) and Meldola's Blue. It has been found to be localized selectively in animal tumors [136] and can retard tumor growth [137,138]. NB has been used as a photosensitizer for oxygen in PDT applications [139,140], in processes that depend on solvent polarity [141,142], as a stain for Escherichia coli in flow cytometry [143], as a DNA probe [144] and many other applications [145,146,147,148]. Due to their high fluorescence quantum yield together with their solvatochromism, they have been used as stains and imaging agents. These dyes present relatively low solubility in aqueous medium as well as their fluorescence is reduced significantly in in the presence of polar medium, which opens up new possibilities to develop aqueous analogues of these benzophenoxazines [149]. Together with the increase of the solubility in water, it is believed that the self-assembly process to form aggregates can be disrupted resulting in an enhancement of the fluorescence intensity [150].

NB shows thermochromic and solvatochromic behavior in its ultraviolet/visible spectra [151]. The variation in the absorption spectrum is due to the equilibrium between the monocation and the neutral molecule, where the monocationic form is the more stable in most solvents. In strong basic conditions the neutral form is observed, where in strong acidic conditions the dicationic and tricationic forms can be observed [152]. The fast decay processes study can be used to get information on the effect of medium condition, basic and acidic, on determining the excited state lifetime on the picosecond scale. It was shown that the reason for the faster decay in acidic conditions results from the formation of dications by reaction of excited state monocations with hydrogen ions [153].

Despite the photophysics of NB in pure solvents is well characterized in literature, the NB interaction with microheterogeneous systems, such as micelles, reverse micelles (RMs) and DNA is still not well understood. Electrochemical studies have shown that NB-DNA duplexes modified microelectrode can be used as a rapid and sensitive method to detect TATA binding to DNA in the presence of other proteins [154]. However, there are no many works done on its interaction with DNA [155,156,157,158]. In a work done on the interaction of NB with biomimicking self-organized assemblies (SDS micelles and AOT reverse micelles) and a genomic DNA (extracted from salmon sperm) (SS DNA), it has been shown that there are two different binding modes of NB with genomic DNA, electrostatic and intercalative modes [144]. There was no explanation for the mechanism related to these interaction modes. The electrostatic mode is believed to be responsible for electron transfer between the probe and DNA, which may result in a quenching process of the NB fluorescence emission intensity when in the presence of low concentration of DNA. The intercalative mode is believed to be the subsequent release of quenching due to the intercalation of the dye in DNA base pairs. In another study, it was shown that binding affinity of the probe is higher with SDS micelles than with the DNAs within its structural integrity in presence of the micelles. The complex rigidity of NB with various DNAs and its fluorescence quenching with DNAs has shown a strong recognition mechanism between NB and DNA [159].

NB was immobilized in two different surfaces, a nonreactive surface (SiO$_2$), with its conduction band at much higher energies, and a reactive surface (SiO$_2$), with a conduction band situated at lower energies. The former is used to directly probe the excited-state dynamics of the dye undisturbed by other competing processes. The latter is used to study the charge injection process from the excited dye into the semiconductor nanocrystallites, acting as an electron acceptor. The transient absorption measurements of NB adsorbed on SiO_2 colloids (inert support) show that the NB aggregates have a relatively short-lived excitonic singlet state (τ = 40 ps) (Table 1). The lifetime of the excited singlet of the monomer in aqueous solution is ~ 390 ps. NB aggregates that were immobilized on reactive surface also inject electrons into SnO_2, resulting in the formation of the the cation radical, $(NB)_2^{\cdot+}$, of the NB aggregates and by the trapping of electrons the in SnO_2 nanocrystallites. The monophotonic dependence of the formation of $(NB)_2^{\cdot+}$ on SnO_2 surface supports the charge transfer from NB aggregates to SnO_2. The rate constant for this heterogeneous electron transfer process is ~ $3.3.10^8$ s^{-1} [160].

4. Aggregation of photosensitizers and its influence in PDT

Most of these dyes form aggregates in the ground state [161,162,163], even when the dye concentration is low (approximately 10^{-6} M) and in the presence of salts and aggregation inducing agents, such as anionic micelles, heparin, polyelectrolyte, liposome and vesicles. The planar structure of such dyes is a key factor that contributes to the approaching and dimerization of the dyes [164,165].

The presence of hydrophobic ligands in the dye structure facilitates the aggregate formation in polar medium. The effects of the planar structure of the dye, hydrophobicity and the interaction with cell membranes were observed in photosynthetic systems II of plants [166] and other systems [167]. Some studies have shown that the interaction among phenothiazines and cyclodextrins results in the aggregate formation with different sizes depending on the cyclodextrins cavity size [162,168].

Studies that were conducted previously have shown that methylene blue molecules form aggregates and the photophysical behavior changes depending on the ground state aggregation. It results in a decreasing of the fluorescence intensity and on the singlet oxygen formation [49]. These studies have shown that the interaction with micelles is responsible for the dimerization process and not the interaction with monomer of the surfactant, as it has been postulated in some works [169]. In this stage of the work it is important to study the nature of the aggregates formed in different negatives interfaces and in biological systems, more specifically in micelles, vesicles and mitochondria.

It is well known that dimerization and medium composition effects changes the energy transfer process among triplets species and molecular oxygen and other triplet suppressors [170,171,172,173,174,175,176,177,178]. Some studies carried out using thionine and MB have shown some of these effects [179]. Azure A, azure B, thionine e MB are dimerized with different dimerization constants.

The aggregation of ionic dyes cannot be assigned to a specific type of chemical interaction. There is a significant contribution of several influences, such as van der Waals interactions, intermolecular hydrogen bounds and pi-electrons interactions, being that, frequently, it is not trivial to evaluate the specific contribution of each one of these interactions [180].

The quantum behavior of extended aggregates of atomic and molecular monomers, containing from a just a few up to thousands of subunits, is attracting increasing attention in chemistry and physics, being that proeminent examples are aggregates of large dye molecules, chromophore assemblies describing the photosynthetic unit of assemblies of ultra-cold atoms [181].

According to their structure, dyes, such as phenotiazinium, exhibit J- or H-aggregates, which present very typical J- or H-absorption bands [182]. The aggregate absorption band is red-shifted in relation to the monomer absorption. These are the J-aggregates showing a very narrow band whose position is well-predicted by a theory ignoring intramolecular vibrations. By contrast, other dyes showed a shift towards the blue (i.e. higher absorption energies) and were termed H-aggregates (hypsochromic shift). Unlike the J-band, the line shape of the H-band generally shows a rich vibrational structure and has a width of the order of the monomeric band [183]. The J-band is polarized parallel to the rods, while the H-band is polarized perpendicularly to the rod long-axis [184].

Self-organized J-aggregates of dye molecules, known for over 60 years, are emerging as remarkably versatile quantum systems with applications in photography, opto-electronics, solar cells, photobiology and as supramolecular fibres [185].

5. Future perspectives

Photodynamic Therapy has been used in clinical applications with significant success. Several studies have focused on the suitable conditions to improve the clinical results, such as the optimization of the incident light intensity. Indeed, the importance of irradiance is a determinant of PDT-induced pain. The increased use of low irradiance PDT may have a considerable impact on pain, which currently is the main limiting factor to successful delivery of PDT in some patients [186].

Another area of improvement of the PDT application is focused on the increase of the aqueous solubility of the photosensitizers. In fact, the photosensitizers require being suitable to several types of administration in biological medium. In this context, interesting photosensitizers that present low water solubility, such as C_{60}, constitute an area of scientific efforts. C60 can be accumulated selectively in the target point. However, the biological application of C60 is limited due to its poor solubility in water [187]. To improve the solubility of C60 in water, several water-soluble derivatives have been synthesized. Furthermore, other solubilization methods for C60 have been explored using cyclodextins, calixerenes, micelles, liposomes, and poly(N-vinyl-2-pyrroridone) (PNVP). In general, core-shell polymer micelles can be formed spontaneously by amphiphilic diblock copolymers due to association between hydrophobic blocks in water. The hydrophobic drugs can be incorporated into the hydrophobic core of the polymer micelle, and thus, the drugs can be solubilized in water. Nanosized water-soluble core-shell type polymer micelles can allow long circulation in the blood stream avoiding reticuloendothelial systems (RESs) and can be utilized for their enhanced permeability and retention (EPR) effect at solid tumor sites.

The production of ROS can be affected by factor, such as the aggregation and photobleaching of the photosensitizer. In fact, photosensitizers such as, for example, magnesium protoporphyrin (MgPpIX), have demonstrated that the aggregation and photobleaching reduce the photodynamic efficiency [188].

Low-level laser therapy has been used to speed up healing process of pressure ulcers due to its antiinflammatory, analgesic, anti-edematous, and scarring effects, since there is no consensus on its effect on infected ulcers [189]. It is an interesting topic to be evaluated in novel studies.

It is known that Gram positive bacteria are more sensitive to PDT as compared to Gram negative species. However, the use of cationic photosensitizers or agents that increase the permeability of the outer membrane allows the effective killing of Gram negative organisms [190]. Some photosensitizers have an innate positive charge, but some approaches are focused on to link photosensitizers to a cationic molecular vehicle, such as poly-L-lysine [190].

Photodynamic therapy has been also applied in dentistry, in endodontic treatments, with auspicious results regarding the control of microbial infections associated to this type of odontologic therapy [191].

The increasing application of PDT has motivated the development of other therapeutic techniques, with similar principles. We can mention the case of Sonodynamic Therapy (SDT). In 1989, Umemura and co-workers first pioneered the development of non-thermal ultrasound activating a group of photosensitizers for treating tumor, which is called Sonodynamic Therapy (SDT) [192]. They reported that the photosensitive compounds activated by ultrasound can kill cancerous cells and suppress the growth of tumor. Otherwise, they also thought highly of that the ultrasound could reach deep-seated tumor and maintain the focus energy in a small volume because of exceedingly strong penetration ability and mature focusing technology [193]. Particularly, SDT was developed from the well-known PDT but only put up low phototoxicity. Therefore, in recent years, along with the lucubration the SDT has attracted considerable attention and has been considered as a promising tumor treatment method [192].

Regarding the development of photosensitizers, it is important to register the relevant role of phthalocyanines. Phthalocyanines (Pcs) are highly delocalized p-conjugated organic systems and exhibit wide variety of roles in a various high technological areas such as semiconductor devices, liquid crystals, sensors, catalysts, non-linear optics, photovoltaic solar cells and PDT [194,195]. They are among the most important promising chemical compounds by advantage of their stability, photophysical, photochemical, redox and coordination properties. The properties of Pcs depend on their molecular composition with the number; position and nature of substituents and type of central metal play an important role in controlling their properties [194].

Indeed, Pcs have many considerable physical and chemical features, which have motivated the interest of several investigators because of their physico-chemical properties [195]. The presence of different substituents on the Pc ring also leads to increased solubility and supramolecular organizations with improved physicochemical characteristics, depending of the interest in terms of application [194]. In fact, phthalocyanines are very versatile chemical systems, which allow great variability of adjustment of properties in the process of chemical synthesis. This great number of structural possibilities has been utilized in many fields, since the different phthalocyanines can be applied in quite different areas, such as gas sensors, semiconductor materials, photovoltaic cells, liquid crystals, optical limiting devices, molecular electronics, non-linear optical applications, Langmuir-Blodgett films, fibrous assemblies and PDT [195].

6. Conclusions

The selection of photosensitizers that are more able to generate an efficient photodynamic action is one of the main questions involving PDT in the present days. The novel generations of photosensitizers is aiming to obtain the maximum quantum yield through the therapeutic window, avoiding spectral ranges that are absorbed by endogenous dyes, such as hemoglobin and melanin. In this way, the previous knowledge regarding the spectroscopic behavior of the new prototypes of photosensitizers is a relevant pre-requisite to the advancement of the types of applications and its respective repercussions, since the efficacy of the methodology depends on the capability of generations of reactive oxygen species (ROS) and reactive nitrogen species (RNS), which are intrinsically related to the optical profile of the photosensitizers.

Acknowledgment

Hueder Paulo Moisés de Oliveira thanks to the financial support propitiated by FAPESP (Project of research support 06/56701-3) and to CNPq to the research grants (479655/2008-1). Thanks to Msc Sandra Cruz dos Santos and Luiza Rosimeri Romano Santin to the revisions. Thanks also FFCLRP-USP, Prof. Amando Siuiti Ito's laboratory.

The author Máira Regina Rodrigues thanks CNPq and FAPESP for the financial support.

Author details

Leonardo M. Moreira
Departamento de Zootecnia (DEZOO), Universidade Federal de São João Del Rei (UFSJ),
São João Del Rei, MG, Brazil

Juliana P. Lyon
Departamento de Ciências Naturais (DCNAT),
Universidade Federal de São João Del Rei (UFSJ),
São João Del Rei, MG, Brazil

Ana Paula Romani
Departamento de Química - Instituto de Ciências Exatas e Biológicas,
Universidade Federal de Ouro Preto, Campus Morro do Cruzeiro, Ouro Preto , MG, Brazil

Phenotiazinium Dyes as Photosensitizers (PS) in Photodynamic Therapy (PDT): Spectroscopic Properties and Photochemical Mechanisms

131

Divinomar Severino
Instituto de Química, Universidade de São Paulo, São Paulo, SP, Brazil

Maira Regina Rodrigues
Universidade Federal Fluminense, Polo Universitário de Rio das Ostras, Rio das Ostras, RJ, Brazil

Hueder P. M. de Oliveira*
*Centro de Ciências Químicas, Farmacêuticas e de Alimentos,
Universidade Federal de Pelotas, Pelotas, RS, Brazil.*

7. References

[1] Raab O (1900) Uber die Wirkung, fluorescirender Stoffe auf infusorien. Z. Biol. 39: 524-546.

[2] Manoto S L, Abrahamse H (2011) Effect of a newly synthesized Zn sulfophthalocyanine derivative on cell morphology, viability, proliferation, and cytotoxicity in a human lung cancer cell line (A549). Lasers Med Sci 26: 523–530.

[3] Machado A E H (2000) Terapia Fotodinâmica: Princípios, potencial de aplicação e perspectivas. Quim. Nova 23: 237-243.

[4] Simplicio F I S, Maionchi F, Hioka N (2002) Photodynamic Therapy: Pharmacological aspects, applications and news from medications development. Quim. Nova 25: 801-807.

[5] Almeida R D, Manadas B J, Carvalho A P, Duarte C B (2004) Intracellular signaling mechanisms in photodynamic therapy. Biochim. Biophys. Acta 1704: 59-86.

[6] Agostinis P, Buytaert E, Breyssens H, Hendrickx N (2004) Regulatory pathways in photodynamic therapy induced apoptosis. Photochem. Photobiol. Sci. 3: 721-729.

[7] Mitton D, Ackroyd R (2008) A brief overview of photodynamic therapy in Europe. Photodiagn. Photodyn. Ther. 5: 103-111.

[8] Dougherty T J, Gomer C J, Henderson B W, Jori G, Kessel D, Korbelik M, Moan J, Peng Q (1998) Photodynamic therapy. J. Natl. Cancer Inst. 90: 889-905.

[9] Tardivo J P, Giglio A D, Paschoal L H, Baptista M S (2006) New photodynamic therapy protocol to treat AIDS-related Kaposi`s sarcoma. Photomed. Laser Sug. 24: 528-531.

[10] Peloi L S, Biondo C E G, Kimura E, Politi M J, Lonardoni M V C, Aristides S M A, Dorea R G C, Hioka N, Silveira T G V (2011) Photodynamic therapy for American cutaneous leishmaniasis: The efficacy of methylene blue in hamsters experimentally infected with Leishmania (Leishmania) amazonensis. Experimental Parasitology 128: 353-356.

[11] Biyani N, Singh A K, Mandal S, Chawla B, Madhubala R (2011) Differential expression of proteins in antimony-susceptible and -resistant isolates of Leishmania donovani. Mol. Biochem. Parasitology 179: 91-99.

[12] Torezan L, Niwa A B M, Neto C F (2009) Photodynamic therapy in dermatology: basic principles and clinical use. An. Bras. Dermatol. 84: 445-459.

* Corresponding Author

[13] Bagazgoitia L, Santos J C, Juarranz A, Jaen P (2011) Photodynamic therapy reduces the histological features of actinic damage and the expression of early oncogenic markers. British Association of Dermatologists 165: 144–151.

[14] Wainwright M, (2007) Phenothiazinium photosensitisers: V. Photobactericidal activities of chromophore-methylated phenothiazinium salts. Dyes and Pigments 73: 7-12.

[15] Prates R A, Kato I T, Ribeiro M S, Tegos G P, Hamblin M R (2011) Influence of multidrug efflux systems on methylene blue-mediated photodynamic inactivation of Candida albicans. J Antimicrob Chemother 66: 1525–1532.

[16] Severino D, Junqueira H C, Gugliotti M, Gabrielli D S (2003) Baptista M S Influence of Negatively Charged Interfaces on the Ground and Excited State Properties of Methylene Blue. Photochem. Photobiol. 77: 459-468.

[17] Balzani V, Scandola F (1991) Supramolecular Photochemistry. Ellis Horwood, West Sussex, UK pp. 89–190.

[18] Kalyanasundaran K. (1987) Photochemistry in Microheterogeneous Systems. Academic Press, Orlando, FL pp. 1–151.

[19] Ghanadzadeh A, Zeini A, Kashef A, Ghanadzadeh A, Zeini A, Kashef A (2007) Environment effect on the electronic absorption spectra of crystal Violet. J. Mol. Liq. 133: 61–67.

[20] Danziger R M, Bareli K H, Weiss K (1967) Laser photolysis of methylene blue. J. Phys. Chem. 71: 2633-2640.

[21] Mellish K J, Cox R D, Vernon D I, Griffiths J, Brown S B (2002) In vitro photodynamic activity of a series of methylene blue analogues. Photochem. Photobiol. 75: 392-397.

[22] Kobayashi M, Maeda Y, Hoshi T, Okubo J, Tanizaki Y (1989) Analysis of the electronic absorption-Spectrum of adsorbed layers of methylene-blue. J. Soc. Dyers Colourists 105: 362-368.

[23] Alarcon E, Edwards A M, Aspee A, Moran F E, Borsarelli C D, Lissi E A, Nilo D G, Poblete H, Scaiano J C (2010) Photophysics and photochemistry of dyes bound to human serum albumin are determined by the dye localization. Photochem. Photobiol. Sci. 9: 93-102.

[24] Jacobs K Y, Schoonheydt R A (1999) Spectroscopy of methylene blue-smectite suspensions. J. Coll. Interf. Sci. 220: 103-111.

[25] Brenneisen P, Wenk J, Redmond R, Wlaschek M, Kochevar I E, Scharffetter-Kochanek K (1999) Requirement for FRAP and P70 ribosomal S6 kinase in the DNA-damage dependent signaling leading to induction of collagenase/MMP-1 and stromelysin-1/MMP-3 after UVB irradiation of dermal fibroblasts. Photochem. Photobiol. 69: 88S-88S.

[26] Schafer H, Stahn R, Schmidt W (1979) Solvent effects on fluorescence quantum yields of thionine and methylene-blue. Zeitschrift Fur Physikalische Chemie-Leipzig 260: 862-874.

[27] Kagan J, Prakash I, Dhawan S N, Jaworski J A (1984) The comparison of several butadiene and thiophene derivatives to 8-methoxypsoralen and methylene-blue as singlet oxygen sensitizers. Photobiochemistry and Photobiophysics 8: 25-33.

[28] Berkoff B, Hogan M, Legrange J, Austin R (1986) Dependence of oxygen quenching of intercalated methylene-blue triplet lifetime on DNA base-pair composition. Biopolymers 25: 307-316.

[29] Gak V Y, Nadtochenko V A, Kiwi J (1998) Triplet-excited dye molecules (eosine and methylene blue) quenching by H_2O_2 in aqueous solutions. J. Photochem. Photobiol. A. Chem 116: 57-62.

[30] Wilkinson F, Helman W P, Ross A B (1993) Quantum yields for the photosensitized formation of the lowest electronically excited singlet-state of molecular-oxygen in solution. J. Phys. Chem. Ref. Data 22: 113-262.

[31] Wainwright M, Phoenix D A, Rice L, Burrow S M, Waring J (1997) Increased cytotoxicity and phototoxity in the methylene blue series via chromophore methylation. J. Photochem. Photobiol. 40: 233-239.

[32] Kamat P V, Lichtin N N (1981) Electron-transfer in the quenching of protonated triplet methylene-blue by ground-state molecules of the dye. J. Phys. Chem. 85: 814-818.

[33] Nilsson R, Kearns D R, Merkel P B (1972) Kinetic properties of triplet-states of methylene-blue and other photosensitizing dyes. Photochem. Photobiol. 16: 109-115.

[34] Grofcsik A, Kubinyil M, Jones W J (1995) Fluorescence decay dynamics of organic dye molecules in solution. J. Mol. Struct. 348: 197-200.

[35] Grofcsik A, Jones W. J. (1992) Stimulated emission cross-sections in fluorescent dye solutions: gain spectra and excited-state lifetimes of Nile blue A and oxazine 720 J. Chem. Sot. Faraday Trans. 88: 1101-1106.

[36] Oliveira H P M, Junior A M, Legendre A O, Gehlen M H (2003) Transferência de energia entre corantes catiônicos em sistemas homogêneos. Quim. Nova 26: 564-569.

[37] Lee S-K, Mills A (2003) Luminescence of Leuco-Thiazine Dyes. J. Fluor. 13: 375-377.

[38] Viswanathan K, Natarajan P (1996) Photophysical properties of thionine and phenosafranine dyes covalently bound to macromolecules. J. Photochem. Photobiol. A. Chem. 95: 245-253.

[39] Tuite E, Kelly J M, Beddard G S, Reid G S (1994) Femtosecond deactivation of thionine singlet states by mononucleotides and polynucleotides. Chem. Phys. Lett. 226: 517-524.

[40] Grofcsik A, Kubinyi M, Jones W J (1996) Intermolecular photoinduced proton transfer in nile blue and oxazine 720. Chem. Phys. Lett. 250: 261-265.

[41] Havelcová M, Kubát P, Nemcová I (2000) Photophysical properties of thiazine dyes in aqueous solution and in micelles. Dyes and Pigments 44: 49-54.

[42] Dutt G B, Doraiswamy S, Periasamy N, Venkataraman B (1990) Rotational reorientation dynamics of polar dye molecular probes by picosecond laser spectroscopic technique. J. Chem. Phys. 93: 8498-8513.

[43] Chen J, Cesario T C, Rentzepis P M (2011) Effect of pH on Methylene Blue Transient States and Kinetics and Bacteria Photoinactivation J. Phys. Chem. A 115: 2702–2707.

[44] Sun H, Hoffman M Z (1993) Protonation of the excited states of ruthenium(II) complexes containing 2,2'-bipyridine, 2,2'-bipyrazine, and 2,2'-bipyrimidine ligands in aqueous solution. J. Phys. Chem. 97: 5014–5018.

[45] Chen J, Cesario T C, Rentzepis P M (2010) Time resolved spectroscopic studies of methylene blue and phenothiazine derivatives used for bacteria inactivation. Chem. Phys. Lett. 498: 81–85.

[46] Acemioglu A, Arık M, Efeoglu H, Onganer Y (2001) Solvent effect on the ground and excited state dipole moments of fluorescein. J. Mol. Struct.: Theochem 548: 165–171.

[47] Svanberg K, Anderson T, Killander D, Wang I, Stenram U, Engels, S A, Berg R, Johansson J, Svanberg S (1994) Photodynamic therapy of nonmelanoma malignant-tumors of the skin using topical delta-amino levulinic acid sensitization and laser irradiation. Br. J. Dermatol 130: 743-751.

[48] Tannock I F, Hill R P (1992) The basic science of oncology, 2nd ed. Mc Graw-Hill, New York.

[49] Dougherty T J, Gomer C J, Henderson B W, Jori G, Kessel D, Korbelik M, Moan J, Peng Q (1998) Photodynamic therapy. J. Natl. Cancer Inst. 90: 889-905.

[50] Ochsner M (1997) Photophysical and photobiological processes in the photodynamic therapy of tumours. J. Photochem. Photobiol. B: Biol, 39: 1-18.

[51] Kochevar I E, Lynch M C, Zhuang S G, Lambert C R (2000) Singlet oxygen, but not oxidizing radicals, induces apoptosis in HL-60 cells. Photochem. Photobiol. 72: 548-553.

[52] Fu Y C, Jin X P, Wei S M, Lin H F, Kacew S (2000) Ultraviolet radiation and reactive oxygen generation as inducers of keratinocyte apoptosis: Protective role of tea polyphenols. J. Toxicol. Environ. Health, Part A 61: 177-188.

[53] Lin C P, Lynch M C, Kochevar I E (2000) Reactive oxidizing species produced near the plasma membrane induce apoptosis in bovine aorta endothelial cells. Exp. Cell Res. 259: 351-359.

[54] Jori G, Fabris C (1998) Relative contributions of apoptosis and random necrosis in tumour response to photodynamic therapy: effect of the chemical structure of Zn(II)-phthalocyanines. J. Photochem. Photobiol. B: Biol. 43: 181-185.

[55] Reddi E, Jori G (1988) Steady-state and time-resolved spectroscopic studies of photodynamic sensitizers - porphyrins and phthalocyanines. Rev. Chem. Interm. 10: 241-268.

[56] Schuitmaker J J, Baas P, Leengoed H L L M v, Meulen F W V, Star W M, Zandwijk N V (1996) Photodynamic therapy: A promising new modality for the treatment of cancer. J. Photochem. Photobiol. B: Biol 34: 3-12.

[57] Lewis L M, Indig G L (2000) Solvent effects on the spectroscopic properties of triarylmethane dyes. Dyes Pigm. 46: 145-154.

[58] Baptista M S, Indig G L (1998) Effect of BSA binding on photophysical and photochemical properties of triarylmethane dyes. J. Phys. Chem B 102: 4678-4688.

[59] Amin K, Baptista M S, Indig G L (1998) Mechanisms of photoinactivation of enzymes mediated by triarylmethane dyes. Biophys. J. 74: 367-367

[60] Indig G L, Bartlett J A, Lewis L M (1999) Effect of self-association and protein finding on the photoreactivity of triarylmethane dyes. Photochem. Photobiol. 69: 78S-78S.

[61] Junqueira H C, Severino D, Dias L G, Gugliotti M S, Baptista M S (2002) Modulation of methylene blue photochemical properties based on adsorption at aqueous micelle interfaces. Phys. Chem. Chem. Phys. 4: 2320-2328.

[62] Dougherty T J, Gomer C J, Henderson B W, Jori G, Kessel D, Korbelik M, Moan J, Peng Q (1998) Photodynamic therapy. J. Natl. Cancer Inst. 90: 889–905.

[63] MacDonald J, Dougherty T J (2001) Basic principles of photodynamic therapy. J. Porphyrins Phthalocyanines 5: 105–129.

[64] Bonnett R. (2000) Chemical Aspects of Photodynamic Therapy. Gordon & Breach, Amsterdam. pp. 1-324.

[65] Dolmans D E, Fukumura D, Jain R K (2003) Photodynamic therapy for cancer. Nat. Rev. Cancer 3: 380–387.

[66] Sharman W M, Allen C M, van Lier J E (1999) Photodynamic therapeutics: basic principles and clinical applications. Drug Discov. Today 4: 507–517.

[67] Phillips D (2010) Light relief: photochemistry and medicine. Photochem. Photobiol. Sci. 9: 1589–1596.

[68] Agostinis P, Berg K, Cengel K A, Foster T H, Girotti A W, Golinick S O, Hahn S M, Hamblin M R, Juzeniene A, Kessel D, Korbelik M, Moan J, Mroz P, Nowis D, Piette J, Wilson B C, Golab J (2011) Photodynamic therapy of cancer: an update. CA Cancer J. Clin. 61: 250–281.

[69] Mitsunaga M, Ogawa M, Kosaka N, Rosenblum L T, Choyke P L, Kobayashi H (2011) Cancer cell-selective in vivo near infrared photoimmunotherapy targeting specific membrane molecules. Nat. Med. 17: 1685–1691.

[70] Gomer C J, Ferrario A, Luna M, Rucker N, Wong S (2006) Photodynamic therapy: combined modality approaches targeting the tumor microenvironment. Lasers Surg Med 38: 516–521.

[71] Moan J, Berg K (1991) The photodegradation of porphyrins in cells can be used to estimate the lifetime of singlet oxygen. Photochem. Photobiol. 53: 549–553.

[72] Ribeiro J N, Jorge R A, Silva A R, Flores A V, Ronchi L M, Tedesco A C (2007) Avaliação da atividade fotodinâmica de porfirinas para uso em terapia fotodinâmica através da fotoxidação de triptofano. Ecl. Quím. 32: 7-14.

[73] Dysart J S, Patterson M S (2005) Characterization of Photofrin photobleaching for singlet oxygen dose estimation during photodynamic therapy of MLL cells in vitro. Phys Med Biol. 50: 2597-2616.

[74] Gorman A, Killoran J, Shea C O, Kenna T, Gallagher W M, Shea D F O (2004) In Vitro Demonstration of the Heavy-Atom Effect for Photodynamic Therapy. J. Am. Chem. Soc. 126: 10619-10631.

[75] Ochsner M (1997) Photophysical and photobiological processes in the photodynamic therapy of tumours. J. Photochem. Photobiol. B: Biol, 39: 1-18.

[76] Verma P, Baldrian P, Gabriel J, Trnka T, Nerud F (2004) Copper–ligand complex for the decolorization of synthetic dyes. Chemosphere 57: 1207–1211.

[77] Wainwright M, Giddens R M (2003) Phenothiazinium photosensitisers: choices in synthesis and application. Dyes Pigm. 57: 245-257.

[78] Tanigushi S, Suzuki N, Masuda M, Hisanaga S, Iwatsubo T, Goedert M, Hasegawa M (2005) Inhibition of heparin-induced tau filament formation by phenothiazines, polyphenols, and porphyrins. J. Biol. Chem. 280: 7614-7623.

[79] Prusiner S B, May B C H, Cohen F E (2004) in: Prion Biology and Diseases (Prusiner, S. B., ed) Cold Spring Harbor LaboratoryPress, Cold Spring Harbor, NY pp. 961–1014.

[80] Rojo J, Picker S M, García J J G, Gathof B S (2006) Inactivación de patógenos en productos sanguíneos. Rev. Med. Hosp. Gen. Mex. 69: 99-107.

[81] Li Y F, Huang C Z, Li M (2002) Study of the interaction of Azur B with DNA and the determination of DNA based on resonance light scattering measurements. Anal. Chim. Acta 452: 285-294.

[82] Tardivo J P, Giglio A D, Paschoal L H C, Ito A S, Baptista M S (2004) Photodiagn. Photodyn. Ther. 1: 345-350.

[83] Peloi L S, Soares R R S, Biondo C E G, Souza V R, Hioka N, Kimura E (2008) Photodynamic effect of light-emitting diode light on cell growth inhibition induced by methylene blue. J. Biosci. 33: 231-237.

[84] Bertolotti S G, Previtali C M (1999) The excited states quenching of phenothiazine dyes by p-benzoquinones in polar solvents. Dyes Pigm. 41: 55-61.

[85] Huang Q, Fu W L, Chen B, Huang J F, Zhang X, Xue Q (2004) Inactivation of dengue virus by methylene blue/narrow bandwidth light system. J. Photochem. Photobiol. B 77: 39-43.

[86] Floyd R A, Schneider J E, Dittme D P (2004) Methylene blue photoinactivation of RNA viruses. Antivir. Res. 61: 141-151.

[87] Wainwright M (2000) Methylene blue derivatives - suitable photoantimicrobials for blood product disinfection? Int. J. Antimicrob. Agents 16: 381-394.

[88] Tardivo J P, Giglio A D, Oliveira C S, Gabrieli D S, Junqueira H C, Tada D B, Severino D, Turchiello R F, Baptista M S (2005) Photodiagnosis Photodyn Ther. 2: 175-191.

[89] J. R. Perussi (2007) Photodynamic inactivation of microorganisms. Quim. Nova, 30, 988 - 994.

[90] Gabrielli D S, Belisle E, Severino D, Kowaltowski A J, Baptista M S (2004) Binding, aggregation and photochemical properties of methylene blue in mitochondrial suspensions. Photochem. Photobiol. 79: 227-232.

[91] Zhao Z, Malinowski E R (1999) Window factor analysis of methylene blue in water. J. Chemom. 13: 83-94.

[92] Heger D, Jirkovsky J, Klán P (2005) Aggregation of Methylene Blue in Frozen Aqueous Solutions Studied by Absorption Spectroscopy. J. Phys. Chem. A 109: 6702-6709.

[93] Bergmann K, O'Konski C T (1963) A spectroscopic study of methylene blue monomer, dimer, and complexes with montmorillonite J. Phys. Chem. 67: 2169-2177.

[94] Moreira L M, Lima A, Soares R R S, Batistela V R, Gerola A P, Hioka N, Bonacin J A, Severino D., Baptista M S, Machado A E H, Rodrigues M R, Codognoto L, Oliveira H P M (2009) Metalloclorophyls of Magnesium, Copper and Zinc: Evaluation of the Influence of the First Coordination Sphere on their Solvatochromism and Aggregation Properties. J. Braz. Chem. Soc. 20: 1653-1658.

[95] Delmarre D, Hioka N, Boch R, Sternberg E, Dolphin D (2001) Aggregation studies of benzoporphyrin derivative. Can. J. Chem. 79: 1068-1074.

[96] Simplicio F I, Maionchi F, Santin O, Hioka N (2004) Small aggregates of benzoporphyrin molecules observed in water-organic solvent mixtures. J. Phys. Org. Chem. 17: 325-331.

[97] Hioka N, Chowdhary R K, Chansarkar N, Delmarre D, Sternberg E, Dolphin D (2002) Studies of a benzoporphyrin derivative with pluronics. Can. J. Chem. 80: 1321-1326.

[98] Tessaro A L, Batistela V R, Gracetto A C, Oliveira H P M, R. Sernaglia R L, Souza V R, Caetano W, Hioka N (2011) Stability of benzoporphyrin photosensitizers in water/ethanol mixtures: pK(a) determination and self-aggregation processes. J. Phys. Org. Chem. 24: 155-161.

[99] Teichert M C, Jones J W, Usacheva M N, Biel M A (2002) Treatment of oral candidiasis with methylene blue-mediated photodynamic therapy in an immunodeficient murine model. Oral Surg Oral Med Oral Pathol Oral Radiol Endod. 93: 155-160.

[100] Almeida J M, Garcia V G, Theodoro L H, Bosco Á F, Nagata M J H, Macarini V C (2006) Photodynamic therapy: an option in periodontal therapy. Arquivos em Odontologia 42: 199-210.

[101] Varma R S, Singh A P (1990) Nucleophilic Addition-Elimination Reactions of 2-Hydrazinobenzothiazoles with Indolin-2,3-diones. J. Indian Chem. Soc. 67: 518-520.

[102] Hashiba I, Ando Y, Kawakami I, Sakota R, Nagano K, Mori T (1979) Jpn. Kokai Tokkyo Koho 79 73,771 1979. (CA 91:P193174v).

[103] Aguiar A, Ferraz A (2007) Fe^{3+}- and Cu^{2+}-reduction by phenol derivatives associated with Azure B degradation in Fenton-like reactions. Chemosphere 66: 947–954.

[104] Archibald F S, (1992) A new assay for lignin-type peroxidases employing the dye Azure B. Appl. Environ. Microbiol. 58: 3110–3116.

[105] Narayana B, Cherian T (2005) A Facile Spectrophotometric Method for the Determination of Periodate Using Azure B. J. Braz. Chem. Soc. 16: 978-981.

[106] Ou C, Yuan R, Chai Y, Tang M, Chai R, He X (2007) A novel amperometric immunosensor based on layer-by-layer assembly of gold nanoparticles–multi-walled carbon nanotubes-thionine multilayer films on polyelectrolyte surface. Anal. Chim. Acta 603: 205–213.

[107] Yang M, Yang Y, Yang Y, Shen G, Yu R (2004) Bienzymatic amperometric biosensor for choline based on mediator thionine in situ electropolymerized within a carbon paste electrode. Anal. Biochem. 334: 127–134.

[108] Huang M, Jiang H, Qu X, Xu Z, Wang Y, Dong S (2005) Small molecules as cross-linkers: fabrication of carbon nanotubes/thionine self-assembled multilayers on amino functionalized surfaces. Chem. Commun. 44: 5560–5562.

[109] Xu Y, Yang L, Ye X, He P, Fang Y (2006) Impedance-Based DNA Biosensor Employing Molecular Beacon DNA as Probe and Thionine as Charge Neutralizer. Electroanalysis 18: 873–881.

[110] Deng L, Wang Y, Shang L, Wen D, Wang F, Dong S (2008) A sensitive NADH and glucose biosensor tuned by visible light based on thionine bridged carbon nanotubes and gold nanoparticles multilayer. Biosens. Bioelectron. 24: 951–957.

[111] Nicotra V E, Mora M F, Iglesias R A, Baruzzi A M (2008) Spectroscopic characterization of thionine species in different media. Dyes Pigm. 76: 315-318.

[112] Muller W, Crothers D M (1975) Interactions of heteroaromatic compounds with nucleic acids. 1. The influence of heteroatoms and polarizability on the base specificity of intercalating ligands. Eur. J. Biochem. 54: 267–277.

[113] Tuite E, Kelly J M (1995) The interaction of methylene blue, azure B, and thionine with DNA: formation of complexes with polynucleotides and mononucleotides as model systems. Biopolymers 35: 419–433.

[114] Long X, Bi S, Tao X, Wang Y, Zhao H (2004) Resonance Rayleigh scattering study of the reaction of nucleic acids with thionine and its analytical application. Spectrochim. Acta Part A: Mol. Biomol. Spectrosc. 60: 455–462.

[115] Jockusch S, Lee D, Turro N J, Leonard E F (1996) Photo-induced inactivation of viruses: adsorption of methylene blue, thionine, and thiopyronine on Qbeta bacteriophage. Proc. Natl. Acad. Sci. U.S.A. 93: 7446–7451.

[116] Rao U, Stec B, Teeter M (1995) Refinement of purothionins reveals solute particles important for latice formation and toxicity. 1. a1-purothionin revisited,. Acta Crystallogr. D. Biol. Crystallogr. D 51: 904–913.

[117] Stec B, Rao U, Teeter M M (1995) Refinement of purothionins reveals solute particles important for lattice formation and toxicity. Part 2: structure of beta-purothionin at 1.7 angstroms resolution. Acta Crystallogr., D Biol. Crystallogr. 51: 914–924.

[118] Johnson K A, Kim E, Teeter M M, Suh S W, Stec B (2005) Crystal structure of alpha-hordothionin at 1.9 Angstrom resolution. FEBS Lett. 579: 2301–2306.

[119] Huang H Y, Wang C M (2010) Phenothiazine: An effective molecular adhesive for protein immobilization. J. Phys. Chem. B 114: 3560–3567.

[120] Paul P, Hossain M, Yadav R C, Suresh Kumar G (2010) Biophysical studies on the base specificity and energetics of the DNA interaction of photoactive dye thionine: spectroscopic and calorimetric approach. Biophys. Chem. 148: 93–103.

[121] Paul P, Kumar G S (2010) Toxic interaction of thionine to deoxyribonucleic acids: Elucidation of the sequence specificity of binding with polynucleotides. J. Hazard. Mater. 184: 620–626.

[122] Göktürk S, Talman R Y (2008) Effect of temperature on the binding and distribution characteristics of thionine in sodium dodecylsulfate micelles. J. Solution Chem. 37: 1709–1723.

[123] Nicotra V E, Mora M F, Iglesias R A, Baruzzi A M (2008) Spectroscopic characterization of thionine species in different media. Dyes Pigm. 76: 315-318.

[124] Lai W C, Dixit N S, Mackay R A (1984) Formation of H aggregates of thionine dye in water. J Phys Chem 88: 5364-5368.

[125] Mackay R A, Gratzel M (1985) The photoreduction of thionine by iron(II) in anionic micelles and microemulsions. Ber Bunsenges Phys Chem 89: 526-530.

[126] Chen S, Yuan R, Chai Y, Xu L, Wang N, Li X, Zhang L (2006) Amperometric hydrogen peroxide biosensor based on the immobilization of horseradish peroxidase (HRP) on the layer-by-layer assembly films of gold colloidal nanoparticles and toluidine blue. Electroanalysis 18: 471–477.

[127] Jiao K, Li Q J, Sun W, Wang Z (2005) Voltammetric detection of the DNA interaction with toluidine blue. Electroanalysis 17: 997–1002.

[128] Tian F, Zhu G (2004) Toluidine blue modified self-assembled silica gel coated gold electrode as biosensor for NADH. Sens. Actuators B 97: 103–108.

[129] Jana A K (2000) Solar cells based on dyes. J. Photochem. Photobiol. A 132: 1–17.

[130] Prento P (2001) A contribution to the theory of biological staining based on the principles for structural organization of biological macromolecules. Biotech. Histochem. 76: 137–161.

[131] Ilanchelian M, Ramaraj R (2011) Binding Interactions of Toluidine Blue O with Escherichia Coli DNA: Formation of Bridged Structure. J. Fluoresc. 21: 1439–1453.

[132] Fei D, Wang X M, Li H B, Ding L S, Hu Y M, Zhang H, Zhao S L (2008) Spectroscopy Studies of Interaction between Methylene Blue and Herring Sperm DNA. Acta Chim. Sin. 66: 443-448.

[133] Arikan B, Tunçay M (2005) Micellar effects and reactant incorporation in reduction of toluidine blue by ascorbic acid. Dyes Pigm. 64: 1-8.

[134] Tuite E, Norden B (1994) Sequence-Specific Interactions of Methylene Blue with Polynucleotides and DNA: A Spectroscopic Study. J. Am. Chem. Soc. 116: 7548-7556.

[135] Liu J, Zou A, Mu B (2010) Toluidine blue: Aggregation properties and distribution behavior in surfactin micelle solution. Colloids Surf., B 75: 496–500.

[136] Staveren H J, Speelman O C, Witjes M J H, Cincotta L, Star W M (2001) Fluorescence imaging and spectroscopy of ethyl Nile blue A in animal models of (Pre)malignancies. Photochem. Photobiol. 73: 32–38.

[137] Morgan J, Potter W R, Oseroff A R (2000) Comparison of photodynamic targets in a carcinoma cell line and its mitochondrial DNA-deficient derivative. Photochem. Photobiol. 71: 747–757.

[138] Singh G, Espiritu M, Shen X Y, Hanlon J G, Rainbow A J (2001) In vitro induction of PDT resistance in HT29, HT1376 and SKN-MC cells by various photosensitizers. Photochem. Photobiol. 73: 651–656.

[139] Lin C W, Shulok J R, Wong Y K, Schambacher C F, Cincotta L, Foley J W (1991) Photosensitization, uptake, and retention of phenoxazine Nile Blue derivatives in human bladder carcinoma cells. Cancer Res. 51: 1109-1116.

[140] Lin C W, Shulok J R (1994) Enhancement of Nile Blue derivative-induced photocytotoxicity by nigericin and low cytoplasmic pH. Photochem. Photobiol. 60: 143-146.

[141] Lee S H, Suh J K, Li M (2003) Determination of bovine serum albumin by its enhancement effect of Nile Blue fluorescence. Bull. Korean Chem. Soc. 24: 45-48.

[142] Krihak M, Murtagh M T, Shahriari M R (1997) A spectroscopic study of the effects of various solvents and sol-gel hosts on the chemical and photochemical properties of Thionin and Nile Blue A. J. Sol-Gel Sci. Technol. 10: 153-163.

[143] Betscheider D, Jose J (2009) Nile Blue A for staining Escherichia coli in flow cytometer experiments. Anal. Biochem. 384: 194-196.

[144] Mitra R K, Sinha S S, Pal S K (2008). Interactions of Nile Blue with Micelles, Reverse Micelles and a Genomic DNA. J. Fluoresc. 18: 423–432.

[145] Lee M H, Lee S W, Kim S H, Kang C (2009) Nanomolar Hg(II) detection using Nile Blue chemodosimeter in biological media. Org. Lett. 11: 2101-2104.

[146] Maliwal B P, Kusba J, Lakowicz J R (1995) Fluorescence energy transfer in one dimension: frequency-domain fluorescence study of DNA-fluorophore complexes. Biopolymers 35: 245-255.

[147] Lakowicz J R, Piszczek G, Kang J S (2001) On the possibility of long-wavelength longlifetime high-quantum-yield luminophores. Anal. Biochem. 288: 62-75.

[148] Tajalli H, Ghanadzadeh Gilani A, Zakerhamidi M S, Tajalli P (2008) The photophysical properties of Nile red and Nile blue in ordered anisotropic media. Dyes Pigm. 78: 15-24.

[149] Jose J, Burgess K (2006) Benzophenoxazine-based fluorescent dyes for labeling biomolecules. Tetrahedron 62: 11021-11037.

[150] Pal M K (1965) Effects of differently hydrophobic solvents on the aggregation of cationic dyes as measured by quenching of fluorescence and/or metachromasia of the dyes. Histochimie 5: 24-31.

[151] Rauf M A, Zaman M Z (1987) Spectral properties of oxazines in various solvents. Spectrochim. Acta A 43: 1171-1172.

[152] Gvishi R, Reisfeld R, Eisen M (1989) Structures, spectra and ground and excited state equilibria of polycations of oxazine-170. Chem. Phys. Letters 161: 455-460.

[153] Grofcsik A, Kubinyi M, Jeremy Jones W (1996) Intermolecular photoinduced proton transfer in nile blue and oxazine 720. Chem. Phys. Letters 250: 261-265.

[154] Gorodetsky A A, Ebrahim A, Barton J K (2008) Electrical detection of TATA binding protein at DNA modified microelectrodes. J. Am. Chem. Soc. 130: 2924–2925.

[155] Chen Q, Li D, Yang H, Zhu Q, Xu J, Zhao Y (1999) Interaction of a novel red-region fluorescent probe, Nile Blue, with DNA and its application to nucleic acids assay. Analyst 124: 901–907.

[156] Huang CZ, Li Y F, Zhang D J, Ao X P (1999) Spectrophotometric study on the supramolecular interactions of nile blue sulphate with nucleic acids. Talanta 49: 495–503.

[157] Yang Y, Hong H Y, Lee I S, Bai D G, Yoo G S, Choi J K (2000) Detection of DNA using a visible dye, Nile Blue, in electrophoresed gels. Anal. Biochem. 280: 322–324.

[158] Ju H, Ye Y, Zhu Y (2005) Interaction between nile blue and immobilized single- or double-stranded DNA and its application in electrochemical recognition. Electrochim. Acta 50: 1361–1367.

[159] Mitra R K, Sinha S S, Maiti S, Pal S K (2009) Sequence dependent ultrafast electron transfer of Nile blue in oligonucleotides. J. Fluoresc. 19: 353–361.

[160] Nasr C, Hotchandani S (2000) Excited-state behavior of Nile blue H-aggregates bound to SiO_2 and SnO_2 colloids. Chem. Mater. 12: 1529-1535.

[161] Bayoumi, A. M. E., Kasha, M. (1959). Exciton-type splitting of eletronics states in hydrogen-bounded molecular dimers of N-heterocyclics. Spectrochim. Acta. 15: 759-760.

[162] Lee C, Sung Y W, Park J W (1999) Multiple equilibria of phenothiazine dyes in aqueous cyclodextrin solutions. J. Phys. Chem. B. 103: 893-898.

[163] Patil K, Pawar R, Talap P (2000) Self-aggregation of methylene blue in aqueous medium and aqueous solutions of Bu₄NBr and urea. Phys. Chem. Chem. Phys. 2: 4313-4317.

[164] Ohline S M, Lee S, Williams S, Chang C (2001) Quantification of methylene blue aggregation on a fused silica surface and resolution of individual absorbance spectra. Chem. Phys. Lett. 346: 9-15.

[165] Zoratti M, Szabò I (1995) The mitochondrial permeability transition. Biochim. Biophys. Acta. 1241: 139-176.

[166] Misran M, Matheus D, Valente P, Hope A (1994) Photochemical electron transfer between methylene blue and quinones. Aust. J. Chem. 47: 209-216.

[167] Collings P J, Gibbs E J, Starr T E, Vafek O, Yee C, Pomerance L A, Pasternack R F (1999) Resonance light scattering and its application in determining the size, shape, and aggregation number for supramolecular assemblies of chromophores. J. Phys. Chem. B 103: 8474-8481.

[168] Liu D, Kamat P V (1996) Dye-capped semiconductor nanoclusters. One-electron reduction and oxidation of thionine and cresyl violet H-aggregates electrostatically bound to SnO2 colloids. Langmuir 12: 2190-2195.

[169] Carroll M K, Unger M A, Leach A M, Morris M J, Ingersoll C M, Bright F V (1999) Interactions between methylene blue and sodium dodecyl sulfate in aqueous solution studied by molecular spectroscopy. Appl. Spect. 53: 780-784.

[170] Sakellarioufargues R, Maurette M T, Oliveros E, Riviere M, Lattes A (1982) Chemical and photochemical reactivity in micellar media and micro-emulsions. 4. Concentration effects on isophorone dimerization. J. Photochem. 18: 101-107.

[171] Sakellarioufargues R, Maurette M T, Oliveros E, Riviere M, Lattes A (1984) Chemical and photochemical reactivity in micellar media and microemulsions. 7. Effect of the interface on the reactivity of excited-states. Tetrahedron 40: 2381-2384.

[172] Reddi E, Jori G, Rodgers M A J, Spikes J D (1983) Flash-photolysis studies of hemato-porphyrins and copro-porphyrins in homogeneous and microheterogeneous aqueous dispersions. Photochem. Photobiol. 38: 639-645.

[173] Oliveros E, Pheulpin P, Braun A M (1987) Comparative-study of the sensitized photooxidation of N-methyl phenothiazine in homogeneous and microheterogeneous media. Tetrahedron 43: 1713-1723.

[174] Daraio M E, Aramendía P F, San Román E A, Braslavsky S E (1991) Carboxylated zinc phthalocyanines. 2. dimerization and singlet molecular-oxygen sensitization in hexadecyltrimethylammonium bromide micelles. Photochem. Photobiol. 54: 367-373.

[175] Kikteva T, Star D, Zhao Z, Baisley T L, Leach G W (1999) Molecular orientation, aggregation, and order in rhodamine films at the fused silica/air interface. J. Phys. Chem. B 103: 1124-1133.

[176] Monte F (1999) Identification of oblique and coplanar inclined fluorescent J-dimers in rhodamine 110 doped sol-gel-glasses. J. Phys. Chem. B 103: 8080-8086.

[177] Monte F, Mackenzie J D, Levy D (2000) Rhodamine fluorescent dimers adsorbed on the porous surface silica gel. Langmuir 16: 7377-7382.

[178] Borba E B, Amaral C L C, Politi M J, Villalobos R, Baptista M S (2000) Photophysical and photochemical properties of pyranine/methyl viologen complexes in solution and in supramolecular aggregates: A switchable complex. Langmuir 16: 5900-5907.

[179] Das S, Kamat P V (1999) Can H-aggregates serve as eight-harvesting antennae? Triplet-triplet energy transfer between excited aggregates and monomer thionine in aerosol-OT solutions. J. Phys. Chem. B. 103: 209-215.

[180] Neumann M G, Gessner F, Cione A P P, Sartori R A, Cavalheiro C C S (2000) Interaction between dyes and clays in aqueous suspension. Quim. Nova 23: 818-824.

[181] Eisfeld A, Schulz G, Briggs J (2011) The influence of geometry on the vibronic spectra of quantum aggregates. J. Lumin. 131: 2555-2564.

[182] Roden J, Eisfeld A, Briggs J S (2008) The J- and H- bands of dye aggregate spectra: Analysis of the coherent exciton scattering (CES) approximation, Chem. Physics 352: 258-266.

[183] Eisfeld A, Briggs J S (2006) The J- and H-bands of organic dyes aggregates, Chem. Physics 324: 376-384.

[184] Eisfeld A, Briggs J S (2007) The Shape of the J-band of pseudoisocyanine, Chem. Phys. Let. 446: 354-358.

[185] Eisfeld A, Briggs J S (2002) The J-band of organic dyes: lineshape and coherence length, Chem. Physics 281: 61-70.

[186] Ibbotson S H (2011) Irradiance is an important determinant of pain experienced during topical photodynamic therapy, J. Am. Acad. Dermatol. 65: 201-202.

[187] Yusa S, Awa S, Ito M, Kawase T, Takada T, Nakashima K, Liu D, Yamago S, Morishima Y (2011) Solubilization of C60 by Micellization with a Thermoresponsive Block Copolymer in Water: Characterization, Singlet Oxygen Generation, and DNA Photocleavage. J. Polym. Sci., Part A: Polym. Chem. 49: 2761–2770.

[188] Ronchi L M, Ribeiro A V F N, da Silva A R, de Sena G L, Jorge R A, Ribeiro J N (2007) The influence of aggregation and photobleaching in the photodynamic activity of magnesium protoporphyrin. Revista Capixaba de Ciência e Tecnologia 2: 5-12.

[189] Benvindo R G, Braun G, de Carvalho A R, Bertolini G R F (2008) Effects of photodynamic therapy and of a sole low-power laser irradiation on bacteria in vitro. Fisioterapia e Pesquisa 15: 53-7.

[190] Demidova T N, Hamblin M R (2004) Photodynamic Therapy targeted to pathogens. International Journal of Immunophatology and Pharmacology 17: 245-254.

[191] Amaral R R, Amorim J C F, Nunes E, Soares J A, Silveira F F (2010) Photodynamic therapy in endodontics - review of literature. RFO, 15: 207-211.

[192] Wang J, Guo Y, Gao J, Jin X, Wang Z, Wang B, Li K, Li Y (2011) Detection and comparison of reactive oxygen species (ROS) generated by chlorophyllin metal (Fe, Mg and Cu) complexes under ultrasonic and visible-light irradiation. Ultrason. Sonochem. 18: 1028–1034.

[193] Tangkuaram T, Wang J, Rodriguez M C, Laocharoensuk R, Veerasai W (2007) Highly stable amplified low-potential electrocatalytic detection of NAD+ at azure-chitosan modified carbon electrodes. Sens. Actuators, B 121: 277–281.

[194] Yuksel F, Durmus M, Ahsen V (2011) Photophysical, photochemical and liquid-crystalline properties of novel gallium(III) phthalocyanines. Dyes Pigm. 90: 191-200.

[195] Dilber G, Durmus M, Kantekin H, Çakır V (2011) Synthesis and characterization of a new soluble metal-free and metallophthalocyanines bearing biphenyl-4-yl methoxy groups. J. Organomet. Chem. 696: 2805-2814.

Physical Spectroscopy

Chapter 6

Atomic and Molecular Low-n Rydberg States in Near Critical Point Fluids*

Luxi Li, Xianbo Shi, Cherice M. Evans and Gary L. Findley

Additional information is available at the end of the chapter

1. Introduction

Since electronic excited states are sensitive to the local fluid environment, dopant electronic transitions are an appropriate probe to study the structure of near critical point fluids (i.e., perturbers). In comparison to valence states, Rydberg states are more sensitive to their environment [1]. However, high-n Rydberg states are usually too sensitive to perturber density fluctuations, which makes these individual dopant states impossible to investigate. (Nevertheless, under the assumption that high-n Rydberg state energies behave similarly to the ionization threshold of the dopant, dopant high-n Rydberg state behavior in supercritical fluids can be probed indirectly by studying the energy of the quasi-free electron, through photoinjection [2–11] and field ionization [12–19].) Low-n Rydberg states, on the other hand, are excellent spectroscopic probes to investigate excited state/fluid interactions.

The study of low-n Rydberg states in dense fluids began with the photoabsorption of alkali metals in rare gas fluids [20, 21]. Later researchers expanded the investigation into rare gas dopants in supercritical rare gas fluids [22–26], and molecular dopants in atomic and molecular perturbers [27–36]. However, none of these previous groups studied dilute solutions near the critical point of the solvent. (The single theoretical study of a low-n Rydberg state in a near critical point fluid was performed by Larrégaray, et al. [35]; this investigation predicted a change in the line shape and in the perturber induced shift of the Rydberg transition.) These results from previous experimental and theoretical investigations of low-n Rydberg states in dense fluids are reviewed in Section 2.

In this Chapter, we present a systematic investigation of the photoabsorption of atomic and molecular dopant low-n Rydberg transitions in near critical point atomic fluids [37–40]. The individual systems probed allowed us to study (dopant/perturber) atomic/atomic interactions (i.e., Xe/Ar) and molecular/atomic interactions (i.e., CH_3I/Ar, CH_3I/Kr, CH_3I/Xe) near the perturber critical point. The experimental techniques and theoretical

---

*This work is adapted from that originally submitted by Luxi Li to the faculty of the Graduate Center of the City University of New York in partial fulfillment of the requirements for the degree of Doctor of Philosophy.

methodology for this extended study of dopant/perturber interactions are discussed in Section 3. Section 4 presents a review of our results for low-n atomic and molecular Rydberg states in atomic supercritical fluids. The accuracy of a semi-classical statistical line shape analysis is demonstrated, and the results are then used to obtain the perturber-induced energy shifts of the primary low-n Rydberg transitions. A striking critical point effect in this energy shift was observed for all of the dopant/perturber systems presented here. A discussion of the ways in which the dopant/perturber interactions influence the perturber-induced energy shift is also presented in Section 4. We conclude with an explanation of the importance of the inclusion of three-body interactions in the line-shape analysis, and with a discussion of how this model changes when confronted with non-spherical perturbers and polar fluids.

2. Perturber effects on low-n Rydberg states

2.1. Supercritical fluids

A supercritical fluid (SCF) exists at a temperature higher than the critical temperature and, therefore, has properties of both a liquid and a gas [41]. For example, SCFs have the large compressibility characteristic of gases, but the potential to solvate materials like a liquid. Moreover, near the critical point the correlation length of perturber molecules becomes unbounded, which induces an increase in local fluid inhomogeneities [41].

The local density $\rho(r)$ of a perturbing fluid around a central species (either the dopant or a single perturber) is defined by [42, 43]

$$\rho(r) = g(r)\rho,$$

where $g(r)$ is the radial distribution function and ρ is the bulk density. For neat fluids and for most dilute dopant/perturber systems, the interactions between the species in the system are attractive in nature. Thus, the perturber forms at least one solvent shell around the dopant (or a central perturber). As the dilute dopant/perturber system approaches the critical density and temperature of the perturber, the intermolecular interactions increase. This increase leads to a change in the behavior of the local density as a function of the bulk density [41]. Therefore, the changes in fluid properties near the critical point are due to the higher correlation between the species in the fluid. These intermolecular correlations are usually probed spectroscopically by dissolving a dopant molecule in the SCF. Since electronic excited states are incredibly sensitive to local fluid environment, fluctuations in the fluid environment can be investigated by monitoring variations in the absorption or emission of the dopant.

Fig. 1 gives three example spectroscopic studies of a dopant in near critical point SCFs. Fig 1a shows the energy position of an anthracene emission line for anthracene doped into near critical point carbon dioxide [44]. Unfortunately, no emission data on non-critical isotherms were measured in this fluorescence emission study. However, the maximum emission position shows a striking critical effect in comparison to a calculated baseline. A more complete investigation of temperature effects on the local density of SCFs via UV photoabsorption [45] of ethyl p-(N,N-dimethylamino)benzoate (DMAEB) in supercritical CHF_3 is presented in Fig. 1b for three isotherms near the critical isotherm. The isotherms shown are at the reduced temperature T_r [where $T_r \equiv T/T_c$ with T_c being the critical temperature of the fluid] of 1.01, 1.06 and 1.11. Although the three isotherms are evenly spaced, the photoabsorption shifts are very similar on the $T_r = 1.06$ and $T_r = 1.11$ isotherms, while the shift on the

$T_r = 1.01$ isotherm is significantly different. Therefore, the perturber induced shift is only temperature sensitive near the critical point of the perturber. Urdahl, *et al.* [46, 47] studied the W(CO)$_6$ T$_{1u}$ asymmetric CO stretching mode doped into supercritical carbon dioxide, ethane and trifluoromethane by IR photoabsorption. All three systems show similar behavior, an example of which is presented in Fig. 1c. Again, the absorption band position changes significantly along the critical isotherm near the critical density. However, extracting the data presented in Fig. 1 is difficult. Moreover, most of the previous work [41] has used dopant valence transitions as probes, and these states tend to be less sensitive to local environments [1]. In the present work, we investigate near critical point SCFs using low-n Rydberg state transitions as the probe. Since Rydberg states are hydrogen-like, it should be possible to model these states within the statistical mechanical theory of spectral line-broadening.

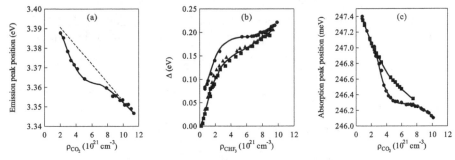

Figure 1. (a) The energy of fluorescence emission for (•) anthracene doped into supercritical carbon dioxide at a reduced temperature $T_r \simeq 1.01$ plotted as a function of carbon dioxide number density ρ_{CO_2}. The dashed line (- - -) is a reference line calculated using perturber bulk densities. Adapted from [44]. (b) The perturber induced energy shifts of the photoabsorption maximum for DMAEB doped into supercritical CHF$_3$ at (•) 30.0°C [i.e., $T_r \simeq 1.01$], (▲) 44.7°C, and (■) 59.6°C, plotted as a function of CHF$_3$ number density ρ_{CHF_3}. Adapted from [45]. (c) The energy position of infrared absorption for the W(CO)$_6$ CO stretching mode in supercritical carbon dioxide at (•) 33°C [i.e., $T_r \simeq 1.01$] and at (■) 50°C, plotted as a function of ρ_{CO_2}. Adapted from [46, 47]. The solid lines in (a) - (c) are provided as a visual aid.

2.2. Theoretical model

Due to the hydrogenic properties of Rydberg states, the optical electron is in general insensitive to the structure of the cationic core. Therefore, both atomic and molecular Rydberg transitions can be modeled within the same theory. In a very dilute dopant/perturber system, assuming that the dopant Rydberg transition is at high energy [i.e. $\beta (E_g - E_e) \gg 1$, $\beta = 1 / (k T)$], the Schrödinger equation is

$$H|\Psi\rangle = E|\Psi\rangle,\tag{1}$$

where the eigenfunction $|\Psi\rangle$ is a product of the dopant electronic wavefunction $|\alpha\rangle$ and the individual perturber wavefunctions $|\psi_i; \alpha\rangle$. The Hamiltonian H in eq. (1) is the sum of several individual Hamiltonians, namely the Hamiltonian for the free dopant, the Hamiltonian for the free perturber, and the Hamiltonian for the dopant/perturber intermolecular correlation. The Hamiltonian H_{FD} for the free dopant is [48]

$$H_{FD} = \sum_{\alpha} E_\alpha |\alpha\rangle\langle\alpha|,\tag{2}$$

where $\alpha = e, g$ with e and g representing the excited state and ground state of the free dopant. The Hamiltonian H_{FP} for the free non-interacting perturber is given by [48]

$$H_{FP} = -\sum_{i}^{N} \frac{\hbar^2}{2m} \nabla_i^2 , \tag{3}$$

where N is the total number of perturbing atoms in the range of the Rydberg optical electron. Finally, the Hamiltonian H_{PD} for the intermolecular interaction is [48]

$$H_{PD} = \sum_{\alpha} \sum_{i} [V_{\alpha}(r_i) + V'(r_i)] |\Psi\rangle\langle\Psi| , \tag{4}$$

where $V(r)$ is the dopant/perturber intermolecular potential, and $V'(r)$ is the perturber/perturber intermolecular potential. Therefore, the total Hamiltonian H is [48]

$$H = H_{FD} + H_{FP} + H_{PD} , \tag{5}$$

which can be rewritten as a ground state Hamiltonian expectation value H_g and an excited state Hamiltonian expectation value H_e.

The absorption coefficient function is given as the Fourier transform [48]

$$\mathfrak{L}(\omega) \equiv \frac{1}{2\pi} \int_{-\infty}^{\infty} dt\, e^{-i\omega t} \langle \vec{\mu}(0) \cdot \vec{\mu}(t) \rangle , \tag{6}$$

where the autocorrelation function (i.e., $\langle \cdots \rangle$) is the thermal average of the scalar product of the dipole moment operator (i.e., $\vec{\mu}$) of the dopant at two different times. This autocorrelation function can be resolved within the Liouville operator formalism to give [48]

$$\langle \vec{\mu}(0) \cdot \vec{\mu}(t) \rangle \equiv \exp\left[\rho_P \langle e^{iL_g t} - 1\rangle_g\right] , \tag{7}$$

where the two-body Liouville operator L_g is defined by

$$L_g \Omega = H_e \Omega - \Omega H_g = [H_g, \Omega] + (E_e - E_g)\Omega , \tag{8}$$

where Ω is an arbitrary operator. However, if lifetime broadening is neglected, only the dopant/perturber interaction and the dopant electronic energy change during the transition. Therefore, the autocorrelation function can be rewritten as [48]

$$\langle \vec{\mu}(0) \cdot \vec{\mu}(t) \rangle \equiv e^{i\omega_0 t} \exp\left[\rho_P \langle e^{i\Delta V t} - 1\rangle_g\right] , \tag{9}$$

where ω_0 is the transition frequency of the neat dopant, ρ_P is the perturber density, and $\Delta V = V_e - V_g$, with V_e and V_g being the excited-state dopant/ground-state perturber and ground-state dopant/ground-state perturber intermolecular potentials, respectively.

In semi-classical line shape theory, the line shape function for an allowed transition is given by [20],

$$\mathfrak{L}(\omega) = \frac{1}{2\pi} \int_{-\infty}^{\infty} dt\, e^{-i[\omega(\mathbf{R}) - \omega_0]t}\, \frac{Z(\beta V_g + it\Delta V)}{Z(\beta V_g)} , \tag{10}$$

where Z is the partition function and $\mathbf{R}$ denotes the collection of all dopant/perturber distances. Under the classical fluid approximation of Percus [49–51], the autocorrelation function $\Phi(t)$ is given by a density expansion [20]

$$\Phi(t) = \ln Z(\beta V_g + it\Delta V) - \ln Z(\beta V_g) = A_1(t) + A_2(t) + \cdots , \qquad (11)$$

where [20, 25]

$$A_n(t) = \frac{1}{n!} \int \cdots \int \prod_{j=1}^{n} d^3 R_j \, \mathfrak{F}(\mathbf{R}_1, \ldots, \mathbf{R}_n)$$

$$\times \prod_{j=1}^{n} \left[\exp(-i\,\Delta V(\mathbf{R}_j)\, t) - 1 \right] . \qquad (12)$$

In eq. (12), $\mathfrak{F}(\mathbf{R}_1, \ldots, \mathbf{R}_n)$ is the Ursell distribution function [25, 26, 49–51], and $\Delta V(\mathbf{R}) = V_e(\mathbf{R}) - V_g(\mathbf{R})$. The Ursell distribution function for two body interactions [25, 26] is $\mathfrak{F}(\mathbf{R}) = \rho_P\, g_{PD}(\mathbf{R})$, where $g_{PD}(\mathbf{R})$ is the perturber/dopant radial distribution function. The three body Ursell distribution function is estimated using the Kirkwood superposition approximation [52] to be

$$\mathfrak{F}(\mathbf{R}_1, \mathbf{R}_2) = \rho_P^2 \left[g_{PP}(|\mathbf{R}_1 - \mathbf{R}_2|) - 1 \right] g_{PD}(\mathbf{R}_1)\, g_{PD}(\mathbf{R}_2) , \qquad (13)$$

where $g_{PP}(\mathbf{R})$ is the perturber/perturber radial distribution function. The density expansion terms are multibody interactions between dopant and perturber over all space. $A_1(t)$ is the dopant/perturber two-body interaction, $A_2(t)$ is the dopant/perturber three-body interaction, and $A_n(t)$ is the dopant/perturber $n+1$ body interaction. Substitution of these Ursell distribution functions into eq. (12) under the assumption of spherically symmetric potentials, gives [25, 26, 37]

$$A_1(t) = 4\pi\rho_P \int_0^\infty dr\, r^2\, g_{PD}(r) \left[e^{-it\Delta V(r)} - 1 \right] , \qquad (14)$$

and

$$A_2(t) = 4\pi\rho_P^2 \int_0^\infty dr_1\, r_1^2\, g_{PD}(r_1) \left[e^{-it\Delta V(r_1)} - 1 \right]$$

$$\times \int_0^\infty dr_2\, r_2^2\, g_{PD}(r_2) \left[e^{-it\Delta V(r_2)} - 1 \right]$$

$$\times \frac{1}{r_1 r_2} \int_{|r_1 - r_2|}^{|r_1 + r_2|} s\left[g_{PP}(s) - 1 \right] ds . \qquad (15)$$

Since the strength of the interaction decreases as the number of bodies involved increases, and since higher order interactions are more difficult to model, most line shape simulations are truncated at the second term $A_2(t)$.

The autocorrelation function can be written as a power series expansion at $t = 0$, namely [21, 48, 53, 54]

$$\Phi(t) = \sum_{n=1}^{\infty} \frac{i^n}{n!} m_n\, t^n , \qquad (16)$$

where the expansion coefficients m_i are given by

$$m_n = \frac{1}{(\sqrt{-1})^n} \frac{d^n}{dt^n} \Phi(t) \Big|_{t=0} . \qquad (17)$$

Using eq. (11) with $A_1(t)$ and $A_2(t)$ from eqs. (14) and (15), the first two expansion coefficients become [25]

$$m_1 = -4\pi\rho_P \int_0^\infty dr\, r^2\, g_{PD}(r)\, \Delta V(r), \tag{18}$$

and

$$m_2 = m_1^2 + 4\pi\rho_P \int_0^\infty dr\, r^2\, g_{PD}(r)\, \Delta V(r)^2$$

$$+ 8\pi^2 \rho_P{}^2 \int_0^\infty \int_0^\infty dr_1\, dr_2\, r_1\, r_2\, g_{PD}(r_1)\, g_{PD}(r_2) \tag{19}$$

$$\times\ \Delta V(r_1)\, \Delta V(r_2) \int_{|r_1-r_2|}^{|r_1+r_2|} s\,[\,g_{PP}(s) - 1\,]\, ds.$$

The above expansion coefficients are equivalent to the moments of the optical coefficient, which are defined as [25]

$$m_n = \int \mathcal{L}(E)\, E^n\, dE. \tag{20}$$

Thus, the perturber induced shift $\Delta(\rho_P)$ in the energy position of the optical coefficient maximum is [25, 26]

$$\Delta(\rho_P) = M_1 = \frac{m_1}{m_0} = \int \mathcal{L}(E)\, E\, dE \Big/ \int \mathcal{L}(E)\, dE, \tag{21}$$

while the full-width-half-maximum of the experimental absorption spectrum is proportional to

$$M_2 = \left[\frac{m_2}{m_0} - M_1^2\right]^{1/2} = \left[\left(\int \mathcal{L}(E)\, E^2\, dE \Big/ \int \mathcal{L}(E)\, dE\right) - M_1^2\right]^{1/2}, \tag{22}$$

where $\mathcal{L}(E)$ is the absorption band for the transition and $E = \hbar(\omega - \omega_0)$ [$\hbar \equiv$ reduced Planck constant].

2.3. Previous studies

The interaction of dopant low-n Rydberg states with dense fluids has previously been the subject of some interest both experimentally and theoretically [1, 22–36]. The first detailed investigation of these interactions was a study of Xe low-n Rydberg states doped into the dense rare gas fluids (i.e., argon, neon and helium) by Messing, et al. [25, 26]. In the same year, Messing, et al. [27, 28] presented their studies of molecular low-n Rydberg states in dense Ar and Kr. A decade later, Morikawa, et al. [34] probed the NO valence and low-n Rydberg state transitions in dense argon and krypton. All of these experiments [25–28, 34] used a basic moment analysis of absorption bands to determine the energy shifts of these bands as a function of perturber number density. The photoabsorption energy shifts were then simulated [25–28, 34] using eq. (18) for various assumptions of intermolecular potentials and radial distribution functions. Messing, et al. also performed line shape simulations under the assumption of a Gaussian line shape [25–28] for selected perturber number densities. As molecular dynamics developed, research groups [30, 31, 33, 35, 36] returned to absorption line shape simulations in an attempt to match the asymmetric broadening observed experimentally.

2.3.1. Xe in Ar, Ne and He

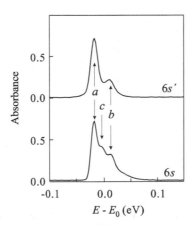

Figure 2. Photoabsorption spectra (relative units) of the Xe $6s$ and $6s'$ transitions doped into Ar at an argon number density of $\rho_{Ar} = 1.47 \times 10^{21}$ cm^{-3} and a temperature of 23.6°C. a corresponds to eq. (23), b to eq. (24), and c to eq. (25). For the $6s$ Rydberg state $E_0 = 8.437$ eV and for the $6s'$ Rydberg state $E_0 = 9.570$ eV. Data from the present work.

When Xe $6s$ and $6s'$ Rydberg state transitions are excited in the presence of low density argon or krypton, satellite bands appear on the higher energy wing of the absorption or emission spectra [22–24]. These blue satellite bands increase with increasing perturber density [22–24] and with decreasing temperature. Therefore, Castex, et al. [22–24] concluded that these blue satellites are caused by the formation of dopant/perturber ground state and excited state dimers. An example of the Xe $6s$ and $6s'$ Rydberg transitions in the presence of argon is shown in Fig. 2. The primary Xe absorption transition, or [37]

$$\text{Xe} + h\nu \xrightarrow{\text{Ar}} \text{Xe}^*, \tag{23}$$

is indicated in Fig. 2 as a. The XeAr dimer transitions that yield the blue satellite bands are [22–24]

$$\text{XeAr} + h\nu \xrightarrow{\text{Ar}} \text{Xe}^* + \text{Ar}, \tag{24}$$

and

$$\text{XeAr} + h\nu \xrightarrow{\text{Ar}} \text{Xe}^*\text{Ar}. \tag{25}$$

These transitions are indicated in Fig. 2 as b and c, respectively.

Detailed investigations [25] of the Xe $6s$ Rydberg state doped in supercritical argon indicated that the energy position at the photoabsorption peak maximum shifted slightly to the red at low argon density and then strongly to the blue (cf. Fig. 3a). Similar results, which are shown in Fig. 3b, were then observed for the Xe $6s'$ Rydberg states in supercritical argon [26]. These studies [25, 26] concluded that both the perturber-induced energy shift and the line shape broadening were temperature independent. However, since the blue satellite bands grow and broaden as a function of perturber number density, the energy position of the maximum absorption (or the first moment from a moment analysis) is not an accurate energy position for the primary Xe Rydberg transition. Thus, modeling the experimental first moment M_1 and

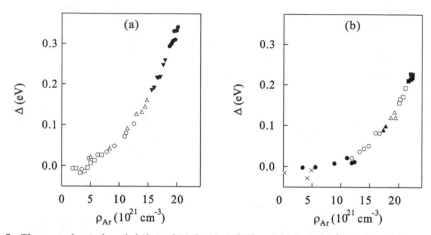

Figure 3. The perturber induced shift Δ of (a) the Xe 6s [25] and (b) the Xe 6s' [26] absorption maximum plotted as a function of argon number density ρ_{Ar} at different temperatures. In (a), the markers are (□) 25°C; (○) −83°C; (△) −118°C; (▼) −138°C; and (●) −163°C. In (b), the markers are for temperature ranges of (×) −23°C to 27°C; (●) −93°C to −63°C; (○) −123°C to −113°C; (▲) −138°C to −128°C; (△) −158°C to −148°C; (□) −173°C to −163°C; and (■) −186°C to −178°C.

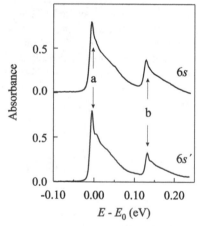

Figure 4. Photoabsorption spectra (relative units) of the CH_3I 6s and 6s' transitions doped into Ar at an argon number density of $\rho_{Ar} = 1.89 \times 10^{21}$ cm^{-3} and a temperature of −79.8°C. a corresponds to eq. (26) and b to eq. (27). For the 6s Rydberg state $E_0 = 6.154$ eV and $E_0 = 6.767$ eV for the 6s' Rydberg state. Data from the present work.

second moment M_2 using eqs. (19) and (20) required three groups of intermolecular potential parameters for different perturber density ranges.

The Xe 6s and 6s' Rydberg transitions [26] in supercritical helium and neon show a similar perturber density dependence as that shown in Fig. 3. However, these two systems do not form ground state or excited state dimers and, therefore, do not have blue satellite bands. Thus, the moment analysis of the photoabsorption spectra presented a more accurate perturber induced shift as a function of perturber number density. Because of the

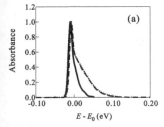 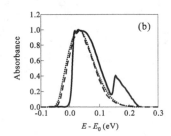 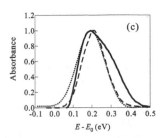

Figure 5. The line shape simulation of the CH_3I 6s transition doped into supercritical argon using (- - -) a semi-classical line shape function data [33] and using ($\cdots$) molecular dynamics [32] in comparison to (—) the photoabsorption spectra at (a) $\rho_{Ar} = 2.0 \times 10^{21}$ cm^{-3}, (b) $\rho_{Ar} = 7.6 \times 10^{21}$ cm^{-3}, and (c) $\rho_{Ar} = 2.0 \times 10^{22}$ cm^{-3}. Experimental data are from the present work.

simplicity of the absorption bands, Messing, et al. [26] simulated the line shapes of the Xe 6s and 6s' transitions in both helium and neon at a single perturber number density using eqs. (10) and (11). These simulations indicated that an accurate line shape could be obtained without blue satellite bands [26]. Unfortunately, no temperature dependence was reported in these papers [25, 26].

2.3.2. CH_3I in Ar and Kr

Since CH_3I is a molecular dopant, vibrational transitions as well as the adiabatic transition appear in the photoabsorption spectra. The adiabatic transition is given by [37]

$$CH_3I + h\nu \longrightarrow CH_3I^*, \tag{26}$$

denoted a in Fig. 4, as well as one quantum of the CH_3 deformation vibrational transition ν_2 in the excited state, or

$$CH_3I + h\nu \longrightarrow CH_3I^* (\nu_2), \tag{27}$$

denoted b, in Fig. 4.

Although vibrational transitions are apparent, CH_3I [27–29, 32] has been investigated extensively because of the "atomic" like nature of the adiabatic 6s and 6s' Rydberg transitions. Messing, et al. [27, 28] extracted the perturber dependent shift $\Delta(\rho_P)$ of the CH_3I 6s and 6s' Rydberg states by performing a moment analysis on the photoabsorption bands using eq. (21). This analysis indicated that $\Delta(\rho_P)$ tended first to lower energy and then to higher energy as ρ_P increased from low density to the density of the triple-point liquid, similar to the trends shown in Fig. 3 for Xe in Ar. However, Messing, et al. [27, 28] did not explore critical temperature effects on $\Delta(\rho_P)$, nor did they correctly account for the vibrational bands on the blue side of the adiabatic Rydberg transition. Messing, et al. [28] did attempt to model the CH_3I 6s Rydberg transition in argon using a semi-classical statistical line shape function under the assumption that the adiabatic and vibrational transitions have exactly the same line shapes, although no comparison between the experimental spectra and the simulated line shapes was provided. Later researchers [30–33] concentrated on the simulation of the CH_3I 6s Rydberg state doped into argon using both molecular dynamics and semi-classical integral methods. Egorov, et al. [33] showed that the semi-classical integral method can yield results comparable to the molecular dynamics calculations of Ziegler, et al. [30–32]. Comparing the semi-classical method and molecular dynamics simulation to experimental spectra (cf. Fig. 5) shows that

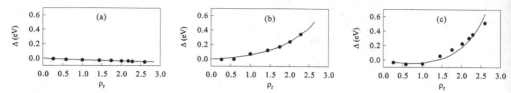

Figure 6. Experimental ($\bullet$) and calculated (—) energy shift of the NO (a) $B'^2\Delta \leftarrow X^2\Pi$ ($v' = 7, 0$) valence transition, (b) $A^2\Sigma^+ \leftarrow X^2\Pi$ ($v' = 0, 0$) Rydberg transition and (c) $C^2\Pi^+ \leftarrow X^2\Pi$ ($v' = 0, 0$) Rydberg transition plotted versus the reduced argon number density [34].

both methods can be used to simulate the experimental spectra with an appropriate choice of intermolecular potentials.

2.3.3. NO in Ar

Morikawa, *et al.* [34] investigated valence and low-n Rydberg transitions doped into supercritical argon. They used eq. (21) to determine the perturber induced shift $\Delta(\rho_P)$ for several low-n Rydberg transitions as well as a valence transition. Under the assumption of spherically symmetric potentials for the ground and excited states of the NO/Ar systems, an accurate moment analysis using eq. (18) was performed (cf. Fig. 6). In The NO valence state transition (cf. Fig. 6a) shows a slight perturber-induced red shift, which differs significantly from the obvious blue shift of the low-n NO Rydberg state transitions shown in Figs. 6b and c. Fig. 6 also shows that low-n Rydberg states make a more sensitive probe to perturber effects than valence transitions. Later groups [35, 36] did line shape simulations to model the experimental spectra. Larrégaray, *et al.* [35] measured the NO $3s\sigma$ transition doped into supercritical argon at selected argon densities. Then, using molecular dynamics they successfully modeled the line shape of the transition. Once the intermolecular potentials and boundary conditions for the molecular dynamics simulations had been set against experimental data, Larrégaray, *et al.* [35] calculated the line shape for NO in Ar at the critical density and temperature. These calculations predicted that the photoabsorption peak maximum position would shift to the blue when argon was near its critical point. Later, Egorov, *et al.* [36] showed that similar results could be obtained using the semi-classical approximation. Therefore, line shape simulations using molecular dynamics and semi-classical line shape theory show comparable results.

3. Experiment techniques and theoretical methodology

3.1. Experimental techniques

All of the photoabsorption measurements presented in Sections 4 and 5 were obtained using the one-meter aluminum Seya-Namioka (Al-SEYA) beamline on bending magnet 8 of the Aladdin storage ring at the University of Wisconsin Synchrotron Radiation Center (SRC) in Stoughton, WI. This beamline, which was decommissioned during Winter 2007, produced monochromatic synchrotron radiation having a resolution of ~ 8 meV in the energy range of 6 - 11 eV. The monochromatic synchrotron radiation enters the vacuum chamber which is equipped with a sample cell (cf. Fig. 7). The photoabsorption signal is detected by a photomultiplier that is connected to the data collection computer via a Keithley 486 picoammeter. The pressure in the vacuum chamber is maintained at low 10^{-8} to high 10^{-9}

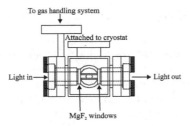

Figure 7. Schematic of the copper experimental cell.

Torr by a Perkin-Elmer ion pump. The experimental copper cell is equipped with entrance and exit MgF_2 windows (with an energy cut-off of 10.9 eV) that are capable of withstanding pressures of up to 100 bar and temperatures of up to 85°C. This cell, which has a path length of 1.0 cm, is connected to an open flow liquid nitrogen cryostat and resistive heater system allowing the temperature to be controlled to within ± 0.5°C with a Lakeshore 330 Autotuning Temperature controller. The gas sample is added through a gas handling system (GHS), which consists of 316-stainless steel components connected by copper gasket sealed flanges. The initial pressure for the GHS and sample cell is in the low 10^{-8} Torr range.

During the initial bakeout, the GHS and the vacuum chamber were heated to 100°C under vacuum for several days to remove any water adsorbed onto the surface of the stainless steel. The initial bake was stopped when the base pressure of the GHS and sample chamber is in the low 10^{-7} or high 10^{-8} Torr range, so that upon cooling the final GHS base pressure was $10^{-8} - 10^{-9}$ Torr. Anytime a system was changed (either the dopant or the perturber), the GHS was again baked in order to return the system to near the starting base pressure. This prevented cross-contamination between dopant/perturber systems.

The intensity of the synchrotron radiation exiting the monochromator was monitored by recording the beam current of the storage ring as well as the photoemission from a nickel mesh situated prior to the sample cell. The light then entered the experimental cell through a MgF_2 window, traveled through the sample and then a second MgF_2 window (cf. Fig. 7) before striking a thin layer of sodium salicylate powder on the inside of a glass window that preceded the photomultiplier tube. An empty cell (acquired for a base pressure of 10^{-7} or 10^{-8} Torr) spectrum was used to correct the dopant absorption spectra for the monochromator flux, for absorption by the MgF_2 windows, and for any fluctuations in the quantum efficiency of the sodium salicylate window.

Two dopants (i.e. methyl iodide and xenon) and five perturbers (i.e. argon, krypton, xenon, carbon tetrafluoride and methane) were investigated. All dopants and perturbers were used without further purification: methyl iodide (Aldrich, 99.45%), argon (Matheson Gas Products, 99.9999%), krypton (Matheson Gas Products, 99.998%), and xenon (Matheson Gas Products, 99.995%). When CH_3I was the dopant, the CH_3I was degassed with three freeze/pump/thaw cycles prior to use. Photoabsorption spectra for each neat dopant and each neat perturber were measured to verify the absence of impurities in the spectral range of interest. The atomic perturber number densities were calculated from the Strobridge equation of state [55] with the parameters obtained from [56] for argon, [57] for krypton, and [58] for xenon. All densities and temperatures were selected to maintain a single phase system in the sample cell.

At temperatures below T_c, a change in density required a change in temperature, since the isotherms are steeply sloped.

The sensitivity of the absorption spectra to local density required that the quality of a data set be monitored by performing basic data analysis during measurements. Any anomalies were corrected by immediately re-measuring the photoabsorption spectrum for the problem density/temperature/pressure after allowing additional time for increased density stabilization. Once a data set was obtained for non-critical temperatures, the photoabsorption data for perturber densities on an isotherm near the critical isotherm were then measured. For the near critical data set we selected a temperature that was $+0.5°C$ above the critical temperature (chosen to prevent liquid formation in the cell during temperature stabilization near the critical density and to minimize critical opalescence during data acquisition). Near the critical density, the consistency of the density step is dependent on the slope of the critical isotherm. If the critical isotherm has a small slope in this region, it becomes difficult to acquire samples at a constant density step size due to our inability to vary the perturber pressure practically by less than 0.01 bar and to the difficulties encountered in maintaining temperature stability. For instance, near the critical density of xenon, a 1 mbar change in pressure or a $0.001°C$ change in temperature causes a density change of 2.0×10^{21} cm^{-3}. Maintaining the necessary temperature stability (i.e., $\pm 0.2°C$) during the acquisition of data along the critical isotherm is difficult with an open flow liquid nitrogen cryostat system and usually required constant monitoring with manual adjustment of the nitrogen flow.

3.2. Theoretical methodology

3.2.1. Line shape function

The experimental line shapes were simulated using the semi-classical statistical line shape function given in eq. (10). Rewriting eq. (10) in terms of the autocorrelation function allows eq. (10) to be given as a Fourier transform, namely [25, 26, 33, 37]

$$\mathfrak{L}(\omega) = \frac{1}{2\pi} \operatorname{Re} \int_{-\infty}^{\infty} dt \, e^{-i\omega t} \langle e^{i\omega(\mathbf{R})t} \rangle , \qquad (28)$$

where $\omega = \omega(\mathbf{R}) - \omega_0$, with ω_0 being the transition frequency for the neat dopant. Eq. (28) neglects lifetime broadening and assumes that the transition dipole moment is independent of $\mathbf{R}$. In the substitution of the exponential density expansion [i.e., eq. (11)] for the autocorrelation function, the general term A_n represents a (n+1)-body interaction [20]. However, since the strength of the interaction decreases as the number of bodies involved increases, and since the higher order interactions are more difficult to model, our line shape simulations are truncated at the second term $A_2(t)$, or three body interactions. Within this approximation, eq. (28) becomes

$$\mathfrak{L}(\omega) = \frac{1}{2\pi} \operatorname{Re} \int_{-\infty}^{\infty} dt \, e^{-i\omega t} \exp[A_1(t) + A_2(t)] , \qquad (29)$$

where the two terms are recalled from eq. (14),

$$A_1(t) = 4\pi \rho_P \int_0^{\infty} dr \, r^2 \, g_{PD}(r) \left[e^{-it \Delta V(r)} - 1 \right] ,$$

and eq. (15),

$$A_2(t) = 4\pi\rho_P^2 \int_0^\infty dr_1\, r_1^2\, g_{PD}(r_1) \left[e^{-it\Delta V(r_1)} - 1 \right]$$
$$\times \int_0^\infty dr_2\, r_2^2\, g_{PD}(r_2) \left[e^{-it\Delta V(r_2)} - 1 \right]$$
$$\times \frac{1}{r_1\, r_2} \int_{|r_1-r_2|}^{|r_1+r_2|} s\left[g_{PP}(s) - 1 \right] ds .$$

The required radial distribution functions (i.c., g_{PP} and g_{PD}) were obtained from the analytical solution of the Ornstein-Zernike equation for a binary system within the Percus-Yevick (PY) closure [59], while the Fourier transform for eq. (29) was performed using a standard fast Fourier transform algorithm [60]. The line shape obtained from the transform of eq. (29) was convoluted with a standard Gaussian slit function to account for the finite resolution ($\sim$ 8 meV) of the monochromator. More detailed discussions are given below.

3.2.2. Fast Fourier transform

A Fourier transform has the general form [60]

$$F(\omega) = \frac{1}{2\pi} \int_{-\infty}^{\infty} f(t)\, e^{-i\omega t}\, dt . \tag{30}$$

Since the line shape function is calculated numerically, the integration limits for eq. (30) must be finite and, therefore, an appropriate integration range must be determined. For any Fourier transformation, the integration limit and the total number of steps are related through [60]

$$\delta t \times \delta \omega = \frac{2\pi}{N} , \tag{31}$$

where δ stands for the sampling interval (i.e., the step size) of the corresponding variable and N is the total number of discrete points. Fourier transforms rely on the fact that data are usually obtained in discrete steps and the generating functions $f(t)$ and $F(\omega)$ can be represented by the set of points

$$f_k \equiv f(t_k), \quad t_k = k\,\delta t, \quad k = 1, \ldots, N,$$
$$F_n \equiv F(\omega_n), \quad \omega_n = n\,\delta\omega, \quad n = -\frac{N}{2}, \ldots, \frac{N}{2} - 1 . \tag{32}$$

Therefore, the function $F(\omega)$ is determined point-wise using

$$F(\omega_n) = \frac{1}{2\pi} \int_{-\infty}^{\infty} f(t)\, e^{-i\omega_n t}\, dt = \frac{1}{2\pi} \sum_{k=1}^{N} f_k\, e^{-i\omega_n t_k}\, \delta t$$
$$= \frac{\delta t}{2\pi} \sum_{k=1}^{N} f_k\, e^{-i2\pi nk/N} . \tag{33}$$

For simplicity, we will define the discrete Fourier transform from time to angular frequency as eq. (33). When computing the Fourier transform from eq. (33), the quickest method is known as a fast Fourier transform (FFT) and requires that the number of steps N be a power of 2. In our calculations, we use a Cooley-Tukey FFT algorithm [60] with $N = 1024$. The requirements for calculating eq. (30) within this FFT algorithm are, therefore, a complex array of the calculated values of the time dependent autocorrelation function truncated to the second term.

Rewriting eq. (29) using Euler's relation yields [38]

$$\langle e^{i\omega(\mathbf{R})t} \rangle = \mathrm{Re}\,\langle e^{i\omega(\mathbf{R})t} \rangle + \mathrm{Im}\,\langle e^{i\omega(\mathbf{R})t} \rangle , \tag{34}$$

where the real and the imaginary parts are given by

$$\mathrm{Re}\,\langle e^{i\omega(\mathbf{R})t} \rangle = \exp\left[\mathrm{Re}\,(A_1(t) + A_2(t))\right]\, \cos\left[\mathrm{Im}\,(A_1(t) + A_2(t))\right] ,$$
$$\mathrm{Im}\,\langle e^{i\omega(\mathbf{R})t} \rangle = \exp\left[\mathrm{Re}\,(A_1(t) + A_2(t))\right]\, \sin\left[\mathrm{Im}\,(A_1(t) + A_2(t))\right] . \tag{35}$$

In eq. (35),

$$\mathrm{Re}[A_1(t) + A_2(t)]$$

$$= 4\pi\rho_P \int_0^\infty dr\, r^2\, g_{PD}(r)\, [\cos(\Delta V(r)\,t) - 1]$$

$$+ 4\pi\rho_P^2 \int_0^\infty \int_0^\infty dr_1\, dr_2\, h(r_1, r_2)\, [\cos(\Delta V(r_1)\,t)\, \cos(\Delta V(r_2)\,t) \tag{36}$$

$$+ 1 - \sin(\Delta V(r_1)\,t)\, \sin(\Delta V(r_2)\,t)$$

$$- \cos(\Delta V(r_1)\,t) - \cos(\Delta V(r_2)\,t)] ,$$

and

$$\mathrm{Im}[A_1(t) + A_2(t)]$$

$$= -4\pi\rho_P \int_0^\infty dr\, r^2\, g_{PD}(r)\, [\sin(\Delta V(r)\,t)]$$

$$- 4\pi\rho_P^2 \int_0^\infty \int_0^\infty dr_1\, dr_2\, h(r_1, r_2)\, [\sin(\Delta V(r_1)\,t)\, \cos(\Delta V(r_2)\,t) \tag{37}$$

$$+ \cos(\Delta V(r_1)\,t)\, \sin(\Delta V(r_2)\,t)$$

$$- \sin(\Delta V(r_1)\,t) - \sin(\Delta V(r_2)\,t)] ,$$

with

$$h(r_1, r_2) = r_1\, g_{PD}(r_1)\, r_2\, g_{PD}(r_2) \int_{|r_1 - r_2|}^{|r_1 + r_2|} s\, [g_{PP}(s) - 1]\, ds .$$

The output of the FFT is a complex function of frequency. The real portion of this complex function is obtained and then convoluted with a standard Gaussian slit function. The final output is the simulated line shape function. Since eqs. (29), (36) and (37) depend on both the radial distribution functions and the ground-state and excited-state intermolecular potentials, these are discussed in more detail below.

3.2.3. Radial distribution function

After significant investigation, we found that the most stable calculation technique for obtaining radial distribution functions for this problem was the analytical solution of the Ornstein-Zernike relation within the Percus-Yevick (PY) closure [59]. Although this solution for a binary system yields four coupled integro-differential equations, dilute solutions (i.e., $\rho_D \ll \rho_P$) allows these equations to be reduced to the calculation of the perturber/dopant radial distribution function $g_{PD}(r)$ and the perturber/perturber radial distribution function $g_{PP}(r)$. This solution is given by [12, 59]

$$g_{PD}(r) = r^{-1} e^{-\beta V_g(r)} Y_{PD}(r)$$

$$g_{PP}(r) = r^{-1} e^{-\beta V'_g(r)} Y_{PP}(r) ,$$

(38)

where

$$Y_{PD}(r) = \int_0^r dt \; \frac{dY_{PD}(t)}{dt} ,$$

(39)

$$Y_{PP}(r) = \int_0^r dt \; \frac{dY_{PP}(t)}{dt} ,$$

with

$$\frac{d}{dr} Y_{PD}(r) = 1 + 2\pi\rho_P \int_0^\infty dt \left(e^{-\beta V_g(t)} - 1 \right) Y_{PD}(t)$$

$$\times \left[e^{-\beta V'_g(r+t)} Y_{PP}(r+t) \right.$$

$$\left. - \frac{r-t}{|r-t|} e^{-\beta V'_g(|r-t|)} Y_{PP}(|r-t|) - 2t \right] ,$$

$$\frac{d}{dr} Y_{PP}(r) = 1 + 2\pi\rho_P \int_0^\infty dt \left(e^{-\beta V'_g(t)} - 1 \right) Y_{PP}(t)$$

$$\times \left[e^{-\beta V'_g(r+t)} Y_{PP}(r+t) \right.$$

$$\left. - \frac{r-t}{|r-t|} e^{-\beta V'_g(|r-t|)} Y_{PP}(|r-t|) - 2t \right] ,$$

(40)

and with V_g and V'_g being the ground state perturber/dopant and ground state perturber/perturber intermolecular potentials, respectively.

3.2.4. Intermolecular potentials

Eqs. (29), (34) - (40) are explicitly dependent on the excited-state and ground-state perturber/dopant intermolecular potentials through $\Delta V(r)$, and are implicitly dependent on the perturber/perturber and perturber/dopant ground-state intermolecular potential via $g_{PP}(r)$ and $g_{PD}(r)$. Thus, these simulations require one to develop a single set of ground-state and excited-state intermolecular potential parameters for each system. A standard Lennard-Jones 6-12 potential, or

$$V(r) = 4\varepsilon \left[\left(\frac{\sigma}{r} \right)^{12} - \left(\frac{\sigma}{r} \right)^{6} \right], \tag{41}$$

was chosen for the atomic perturber/perturber ground-state intermolecular interactions, and non-polar dopant/perturber ground-state intermolecular interactions. The ground-state molecular perturber/perturber intermolecular potential was a two-Yukawa potential, or

$$V(r) = -\frac{\kappa_0 \varepsilon}{r} \left[e^{-z_1 (r - \sigma)} - e^{-z_2 (r - \sigma)} \right]. \tag{42}$$

The ground-state polar dopant/perturber intermolecular interactions were modeled using a modified Stockmeyer potential

$$V(r) = 4\varepsilon' \left[\left(\frac{\sigma'}{r} \right)^{12} - \left(\frac{\sigma'}{r} \right)^{6} \right] - \frac{1}{r^6} \alpha_P \mu_D^2, \tag{43}$$

which can be rewritten in standard Lennard-Jones 6-12 potential form [12], with

$$\varepsilon = \varepsilon' \left[1 + \frac{\alpha_P \mu_D^2}{4 \varepsilon' \sigma'^6} \right]^2,$$

$$\sigma = \sigma' \left[1 + \frac{\alpha_P \mu_D^2}{4 \varepsilon' \sigma'^6} \right]^{-1/6}.$$

(The modified Stockmeyer potential includes orientational effects via an angle average that presumes the free rotation of the polar dopant molecule.) An exponential-6 potential, given by

$$V(r) = \frac{\varepsilon}{1 - (6/\gamma)} \left\{ \frac{6}{\gamma} e^{\gamma(1-\chi)} - \chi^{-6} \right\}, \tag{44}$$

was chosen for the excited-state dopant/ground-state perturber interactions. In eqs. (41) - (44), ε is the well depth, σ is the collision parameter, α_P is the perturber polarizability, μ_D is dopant dipole moment, $\chi \equiv r/r_e$ (where r_e is the equilibrium distance), and γ is the potential steepness.

The Lennard-Jones parameters for the atomic fluids and the modified Stockmeyer parameters for the CH_3I/Ar and CH_3I/Kr interactions were identical to the parameters used to model accurately the perturber-induced shift of the dopant ionization energy for methyl iodide in argon [12–14, 61], krypton [12, 15, 61] and xenon [16, 61]. The parameters κ_0, z_1, z_2, ε and σ in eq. (42) were adjusted to give the best fit to the phase diagram of the perturber [61].The parameters ε, σ, χ and γ in eq. (44) were adjusted by hand to give the "best" fit to the experimental absorption spectra of the dopant low-n Rydberg states in each of the fluids investigated here.

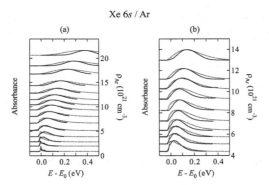

Figure 8. Selected photoabsorption spectra (–, relative scale) and simulated line shapes ($\cdots$) for the Xe 6s Rydberg transitions at (a) non-critical temperatures and (b) on an isotherm (i.e., $-121.8°C$) near the critical isotherm. The data are offset vertically by the argon number density ρ_{Ar}. The transition energy is $E_0 = 8.424$ eV for the unperturbed Xe 6s Rydberg transition.

4. Atomic perturbers

4.1. Xe low-n Rydberg states in Ar

4.1.1. Xe absorption

The Xe 6s and 6s' Rydberg states (where s and s' denote the $J = 3/2$ and $J = 1/2$ angular momentum core state, respectively) were experimentally measured in dense argon. As is discussed in Section 2.4.1, when Xe interacts with Ar, ground and excited state dimers form. These dimers are evidenced by blue satellite bands that arise on the higher energy side of the primary Rydberg transition. The Xe 6s Rydberg transition has two such blue satellite bands corresponding to eqs. (24)-(25), whereas the Xe 6s' Rydberg transition has a single blue satellite band corresponding to eq. (24) (cf. Fig. 2) [22–24]. The absence of the ground state XeAr dimer to XeAr eximer transition [i.e., eq. (25)] for the Xe 6s' Rydberg state may be caused by an extremely short lifetime preventing our ability to detect the transition or by the XeAr eximer decomposing during the excitation.

The solid lines in Fig. 8 present selected experimental Xe 6s Rydberg transitions doped into supercrtical argon at non-critical temperatures and along an isotherm near the critical isotherm offset by the argon number density. (Similar data for the Xe 6s' Rydberg transitions are not shown for brevity.) It can be clearly seen that the Rydberg transitions broaden as a function of the argon number density. The maximum of the absorption band also shifts first slightly to the red and then strongly to the blue, similar to the original observations of Messing, *et al.* [25, 26]. Since the ground state interaction between Xe and Ar (or XeAr and Ar) is attractive, the ground states are stabilized by the argon solvent shell. The slight red shift observed at low argon number densities indicates that the xenon excited states (either Xe* or Xe*Ar) are also stabilized by the argon solvent shell. As the density increases, however, argon begins to shield the optical electron from the xenon cationic core, thereby decreasing the binding energy of the optical electron . Thus, as the density of argon increases the energy of the excited state also increases, leading to a blue shift in the transition energy at higher densities. Although not shown, the overall blue shift of the 6s Rydberg transition band is much larger than that of the 6s' band at the triple point liquid density of argon. This difference in overall shift is caused by the difference in the core state of the cation, since the $J = 1/2$ core state has a permanent quadrupole moment. This permanent quadrupole moment increases the interaction of the cationic core and the optical electron, thereby implying that the optical electron is less perturbed by the argon solvent shell.

However, since the blue satellite bands also broaden and shift with increasing argon density, the primary Xe transition becomes indistinguishable at medium to large argon number densities. Thus, the argon-induced energy shift of the primary Xe transition cannot be investigated directly using these data. Therefore, to probe perturber critical effects on the dopant excited states, we must first accurately simulate the absorption spectra over the entire argon density range at non-critical temperatures and on an isotherm near the critical isotherm.

4.1.2. Discussion

In order to simulate accurately the absorption spectra at high density, any line shape simulation has to include the primary transition, denoted a in Fig. 2 and given by eq. (23), as well as the two XeAr dimer transitions that yield the blue satellite bands, denoted b and c in Fig. 2 and given by eqs. (24) and (25), respectively. For the simulation of Xe in Ar, we chose to use eq. (41) for the ground state Ar/Ar, Xe/Ar, and XeAr/Ar interactions and eq. (44) for the Xe*/Ar and Xe*Ar/Ar interactions. We also required that the simulation use a single set of intermolecular potential parameters for the entire argon density range at non-critical temperatures and along the critical isotherm. All intermolecular potential parameters except the Ar/Ar ground state parameters were adjusted by hand to give the best simulated line shape in comparison to the experimental data. The values of these parameters are given here in Appendix A [37, 40].

The relative intensities of the simulated bands were set by comparison to the absorption spectra of Xe doped into argon at argon number densities where all bands could be clearly identified. Experimentally, at low argon number densities, the ratio of heights between the b band and the primary transition is 0.2 for both the Xe $6s$ and $6s'$ Rydberg states in Ar. For the Xe $6s$ Rydberg state in Ar, the ratio of heights between the c band and the primary transition is 0.45. Although for concentrated Xe systems, the ratio of heights for the blue satellite bands to the primary transition would increase with decreasing temperature or increasing perturber number density, this is not the case for the very dilute Xe/Ar system investigated here (i.e., [Xe] < 10 ppm for all argon number densities). Therefore, we can assume that the intensity ratio of the blue satellite bands to the primary transitions stays constant at different temperatures and different argon densities.

The dotted lines in Fig. 8 are the simulated line shapes for the Xe $6s$ transition at non-critical temperatures (cf. Fig. 8a) and on an isotherm ($-121.8°C$) near the critical isotherm (cf. Fig. 8b). A similar figure for the Xe $6s'$ transition is not shown for brevity. Clearly, the simulated spectra closely match the experimental spectra for all densities. Both the simulated and experimental line shapes show a slight red shift at low argon number densities, followed by a strong blue shift at high argon number densities. With these accurate line shape simulations, moment analyses can be performed on the primary transition in order to investigate perturber critical point effects, as well as to discuss trends in solvation of different dopant electronic transitions in the same simple atomic fluid.

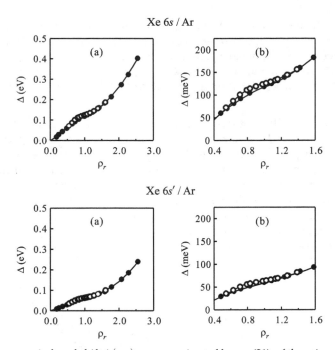

Figure 9. (a) The argon induced shift $\Delta(\rho_{Ar})$, as approximated by eq. (21), of the primary transition for the Xe 6s and 6s' Rydberg states as a function of argon number density ρ_{Ar} at (•) non-critical temperatures and (○) along an isotherm near the critical isotherm. (b) An expanded view of $\Delta(\rho_{Ar})$ The solid lines are a visual aid. See text for discussion.

A line shape analysis was performed on the accurate simulations of the primary Xe 6s and 6s' Rydberg transitions in order to determine the average argon induced shift $\Delta(\rho_{Ar})$ of the primary transition, as approximated from the first moment [i.e., eq. (21)]. This moment analysis is shown in Fig 9 as a function of reduced argon number density ρ_r, where $\rho_r = \rho_{Ar}/\rho_c$ with $\rho_c \equiv 8.076 \times 10^{21}$ cm^{-3} [56]. Fig 9b shows an enhanced view of the perturber critical region with a critical effect in $\Delta(\rho_{Ar})$ clearly apparent. The absence of the red shift observed by Messing, et al. [25, 26] (cf. Fig. 3) results from our performance of a moment analysis on a blue degraded band, instead of a direct non-linear least square analysis using a Gaussian fit function on the primary transition. In other words, while the peak of the primary transition red shifts slightly at low argon densities, the first moment of the band does not, due to the perturber induced broadening.

General trends emerged in the behavior of the simulated line shape as a function of the intermolecular potential parameters. For instance, we observed that the strength of the asymmetric blue broadening of a band increases with increasing $\Delta r_e \equiv r_e^{(g)} - r_e^{(e)}$ [where $r_e^{(i)}$ is the equilibrium dopant/perturber distance for either the ground state dopant ($i = g$) or the excited state dopant ($i = e$)]. However, the overall perturber-induced energy shift of the band depended on the ground state intermolecular potential well depth $\varepsilon^{(g)}$ as well as Δr_e. The slight red shift at low perturber number densities, however, was controlled by the excited state intermolecular potential well depth $\varepsilon^{(e)}$. Comparison of the Xe 6s and 6s' transition in Ar shows that the 6s Rydberg state broadens and shifts to higher energies more quickly than does the 6s' state. Since both transitions are excited from the same ground state (implying that the ground state intermolecular potential parameters remain unchanged), Δr_e must decrease and $\varepsilon^{(e)}$ increase in order to simulate the Xe 6s' Rydberg state in argon correctly. These general trends proved helpful when determining the intermolecular potential parameters for new systems.

Messing, et al. [25, 26] concluded that the argon induced energy shift is density dependent and temperature independent. However, both our experimental absorption spectra and the line shape simulations show a distinct temperature dependence near the argon critical point. To test the sensitivity of the perturber critical point effect, we extracted the perturber dependent shift $\Delta(\rho_{Ar})$ of the simulated primary Xe 6s transition in supercritical argon near the critical density along three different isotherms (i.e., $T_r = 1.01$, 1.06 and 1.11, where $T_r = T/T_c$ with $T_c = -122.3°C$). These data are shown in Fig. 10a and clearly indicate that the critical effect is extremely sensitive to temperature and can be easily missed if the temperature of the system is not maintained close to the critical isotherm.

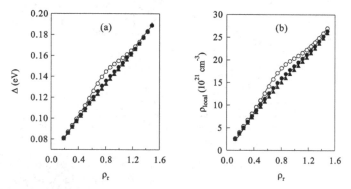

Figure 10. (a) The calculated argon induced shift $\Delta(\rho_{Ar})$ doped into supercritical argon plotted as a function of reduced argon number density at a reduced temperature $T_r = 1.01$ (o), 1.06 (•) and 1.11 (▲). (b) The local densities ($\rho_{local} = g_{max}\,\rho_{bulk}$) of the first argon solvent shell around a central Xe atom plotted as a function of reduced argon number density at a reduced temperature $T_r \simeq 1.01$ (o), 1.06 (•) and 1.11 (▲). The solid lines are provided as visual aid.

If we return to the line shape equation [i.e., eq. (29)], we observe that the two-body interaction term $A_1(t)$ and the three-body interaction term $A_2(t)$ depend on the difference between the excited state and ground state intermolecular potentials and on the perturber/dopant radial distribution function. Since the potential difference will not depend dramatically on temperature, the critical point effect must be dominated by changes in the perturber/dopant radial distribution function $g_{PD}(r)$. In Fig. 10b, we plot the local density of the first solvent shell as a function of the bulk reduced argon number density on the same three isotherms. The $T_r = 1.01$ isotherm shows a much larger density deviation near the critical density in comparison to the other two isotherms. Thus, the argon induced blue shift is caused by the first perturber shell shielding the cationic core from the optical election. This increase in shielding decreases the binding energy of the electron, thereby increasing the excitation energy.

4.2. CH$_3$I low-n Rydberg states in Ar, Kr and Xe

4.2.1. CH$_3$I absorption

The CH$_3$I 6s and 6s' Rydberg states doped into supercritical argon, krypton and xenon were investigated both experimentally and theoretically [38, 40] from low perturber number density to the density of the triple point liquid, at both non-critical temperatures and on an isotherm near (i.e., $+0.5°C$) the critical isotherm of the perturber. The CH$_3$I 6s and 6s' Rydberg states show perturber-induced energy shifts and broadening similar to that observed for the Xe low-n Rydberg states in supercritical argon. The peak positions of the absorption spectra shift to the red slightly and then strongly to the blue as a function of perturber number densities. This is similar to the behavior for CH$_3$I in dense rare gases observed by Messing, *et al.* [27, 28]. Unlike Xe, which forms heterogenous dimers in argon, the CH$_3$I/perturber interactions are weaker. Thus, CH$_3$I does not possess blue satellite bands caused by dimer or excimer formation. However, CH$_3$I does possess a strong vibrational transition on the blue side of the adiabatic transition. Fig. 4 shows the absorption of both the 6s and 6s' Rydberg states of CH$_3$I and clearly illustrates the vibrational state, which represents the CH$_3$ group deformation vibrational band ν_2. The solid lines in Figs. 11 - 13 represent selected photoabsorption spectra for the CH$_3$I 6s Rydberg transition doped into supercrtical argon, krypton and xenon, while similar plots for the CH$_3$I 6s' transition are not shown for brevity. Experimental spectra of CH$_3$I in Xe at number densities between 5.0×10^{21} cm^{-3} and 7.0×10^{21} cm^{-3} could not be obtained, because of the large density deviation induced by small temperature fluctuations ($\approx 2.0 \times 10^{21}$ cm^{-3} for a $0.001°C$ temperature change) in this density region.

The experimental absorption of CH$_3$I low-n Rydberg transitions shows that as the perturber number density increases, the ν_2 vibrational band broadens and shifts until it merges with the adiabatic transition. Therefore, determining the perturber induced shift $\Delta(\rho_P)$ of the adiabatic transition from a simple moment analysis of the spectra presented in Figs. 11 - 13 is not possible, and we must perform an accurate line shape analysis of these data in order to extract $\Delta(\rho_P)$ and investigate the perturber critical effect. However, some qualitative information can be gleaned from Figs. 11 - 13. First, the rate of the broadening and the rate of shift for both the adiabatic transition band and the ν_2 vibrational transition band differ dramatically for different perturbers. However, although not shown, the CH$_3$I 6s and 6s' transitions have almost the same perturber induced shift, which differs from the behavior observed for the Xe in Ar system previously presented.

4.2.2. Discussion

Although CH$_3$I in the rare gases does not form dimers or excimers, the accurate simulation of the low-n Rydberg transitions must include both the adiabatic transition, given by eq. (26) and denoted a in Fig. 4, as well as one quantum of the CH$_3$ deformation vibrational transition ν_2 in the excited state, given by eq. (27) and denoted b in Fig. 4. For all of the simulations presented here, we again chose eq. (41) for the ground-state perturber/perturber intermolecular interactions. All of the ground-state dopant/perturber interactions, on the other hand, were approximated with eq. (43). The excited-state dopant/ground state perturber interactions were again modeled using eq. (44). All intermolecular potential parameters except the Ar/Ar, Kr/Kr, Xe/Xe, CH$_3$I/Ar, and CH$_3$I/Kr ground state potential parameters were adjusted by hand to give the best simulated line shape in comparison to our experimental absorption spectra. (The Ar/Ar, Kr/Kr, Xe/Xe, CH_3I/Ar, and CH_3I/Kr ground-state potential parameters used are in accord with those employed in our earlier studies of the quasi-free electron energy in rare gas perturbers [12].) Appendix A gives the values for all intermolecular potential parameters used in the line shape simulations presented here. The relative intensities of the simulated bands were fixed by comparison to the absorption spectra of CH_3I at perturber number densities where all bands (i.e., the adiabatic and vibrational transitions) could be clearly identified. Experimentally, at low perturber number densities the ratio of the vibrational band intensity to the adiabatic transition intensity is 0.22 for both the CH_3I $6s$ and $6s'$ Rydberg states in all three perturbers.

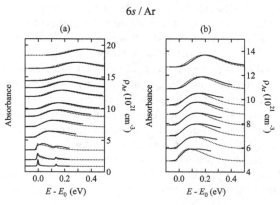

Figure 11. Selected photoabsorption spectra (—, relative units) and simulated line shapes ($\cdots$) for the CH_3I $6s$ Rydberg transition in argon at (a) non-critical temperatures and (b) on an isotherm ($-121.8°C$) near the critical isotherm. The data are offset vertically by the argon number density ρ_{Ar}. The transition energy is $E_0 = 6.154$ eV for the unperturbed CH_3I $6s$ Rydberg transition. The variation between experiment and simulation is caused by other vibrational transitions and by perturber-dependent lifetime broadening not modeled here.

The dotted lines in Figs.11 - 13 present the simulated line shapes (dotted lines) of the low-n CH_3I Rydberg transitions in the atomic perturbers at non-critical temperatures and on an isotherm near the critical isotherm of the perturber. As was true for Xe in Ar, the simulated spectra closely match the experimental spectra for all densities. Both the simulated and experimental line shapes show a slight red shift at low perturber number densities, followed by a strong blue shift at high perturber densities. Given the accuracy of the simulated line shapes, simulated spectra for CH_3I in Xe in the region where experimental data were unobtainable are also presented in Fig. 13. We should note here that we were able to model the CH_3I $6s$ and $6s'$ Rydberg states in Ar using the same set of intermolecular potential parameters for both states. This behavior was also observed for the CH_3I $6s$ and $6s'$ Rydberg states in Kr. With identical potential parameters, the perturber induced shift $\Delta(\rho_P)$ will be the same for the $6s$ and $6s'$ states. The independence of $\Delta(\rho_P)$ on the dopant cationic core state is different from that observed for Xe low-n Rydberg states in Ar and will be discussed in more detail below. The accurate line shape simulations allow $\Delta(\rho_P)$ for the adiabatic transitions to be extracted using eq. (21).

As with Xe in Ar, the accurate line shape simulations allow a moment analysis to be performed on the CH_3I low-n adiabatic Rydberg transition to obtain the perturber induced shift $\Delta(\rho_P)$

6s / Kr

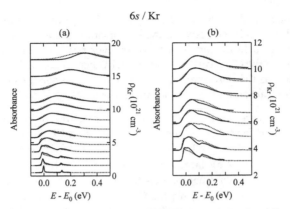

Figure 12. Selected photoabsorption spectra (—, relative units) and simulated line shapes ($\cdots$) for the CH_3I 6s Rydberg transition in krypton at (a) non-critical temperatures and (b) on an isotherm ($-63.3°C$) near the critical isotherm. The data are offset vertically by the krypton number density ρ_{Kr}. The transition energy is $E_0 = 6.154$ eV for the unperturbed CH_3I 6s Rydberg transition. The variation between experiment and simulation is caused by other vibrational transitions and by perturber-dependent lifetime broadening not modeled here.

6s / Xe

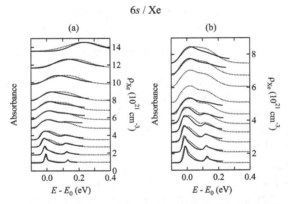

Figure 13. Selected photoabsorption spectra (—, relative units) and simulated line shapes ($\cdots$) for the CH_3I 6s Rydberg transition in xenon at (a) non-critical temperatures and (b) on an isotherm ($17.0°C$) near the critical isotherm. The data are offset vertically by the xenon number density ρ_{Xe}. The transition energy is $E_0 = 6.154$ eV for the unperturbed CH_3I 6s Rydberg transition. The variation between experiment and simulation is caused by other vibrational transitions and by perturber-dependent lifetime broadening not modeled here.

from eq.(21). The first moment of the simulated CH_3I 6s adiabatic transition is plotted as a function of the reduced perturber number density ρ_r in Fig. 14 for the 6s transition. (A similar figure for the 6s′ transition is not shown for brevity.) The first moment of the simulated adiabatic band does not red shift at low perturber density, as was originally stated by Messing, *et al.* [27, 28]. This absence of a red shift is again caused by the blue degradation of the adiabatic transition, which places the average energy (i.e., the first moment) of the band to the high energy side of the absorption maximum. The ground state interaction between CH_3I and the perturber is attractive, and therefore the ground state of the dopant is stabilized by the perturber solvent shell. The slight red shift of the absorption maximum observed at low perturber number densities is indicative of the stabilization of the CH_3I excited states by the perturber solvent shell. As the density increases, however, perturber molecules begin to shield the optical electron from the CH_3I cationic core, thereby increasing the excitation energy of the optical electron. Thus, as the perturber density increases, the energy of the excited state also increases, leading to a blue shift at higher perturber densities.

The 6s and 6s' Rydberg states correspond to an optical electron in the same Rydberg orbital, but with the cation in a different core state: $J = 3/2$ for s and $J = 1/2$ for s', where J is the total angular momentum of the core. In our investigation of $\Delta(\rho_P)$ for Xe in Ar, we found that $\Delta(\rho_P)$ of the 6s transition is 0.2 eV larger than that for the 6s' transition, indicating that the change in the core quadrupole moment affects the dopant/perturber interactions in a dense perturbing medium. However, $\Delta(\rho_P)$ for the CH_3I 6s and 6s' Rydberg transitions near the triple point density are identical to within experimental error for the perturbers argon and krypton, and differ only slightly (i.e., 30 meV) for CH_3I in xenon. The insensitivity of these CH_3I/perturber systems to the change in the CH_3I cationic core is probably caused by the large permanent dipole moment of CH_3I, which masks the effect of the quadrupole moment. Xenon, however, is extremely sensitive to electric fields because of its large polarizability. Therefore, the slight difference between the xenon induced shifts of the CH_3I 6s and 6s' Rydberg transitions may well be caused by small changes in the permanent dipole moment of CH_3I influencing changes in the induced dipole or local quadrupoles in the xenon perturber.

A critical point effect on the 6s and 6s' transition energies is also apparent in Fig. 14 for all three perturbers. The CH_3I 6s adiabatic transition in argon is blue-shifted by 20 meV near the critical temperature and critical density, while those in krypton and xenon are blue-shifted by 30 meV and 15 meV, respectively. Identical results are obtained for the CH_3I 6s' adiabatic transitions in argon and krypton. However, a smaller critical effect of 5 meV is observed for the CH_3I 6s' transition in xenon, which is related to the smaller overall blue shift of the CH_3I 6s' transition in comparison to the 6s transition.

In the low to medium density range, the energy of the absorption maximum for the 6s and 6s' CH_3I Rydberg states has a larger red shift in xenon, which is caused by the larger xenon polarizability. The CH_3I Rydberg states also broaden more quickly in xenon. This increased broadening is probably due to a combination of increased xenon polarizability and an increase in the probability of collisional de-excitation due to the size of xenon. However, $\Delta(\rho_P)$ is larger for argon than for krypton and xenon. This change is caused by an overall decrease in the total number of perturber atoms within the first solvent shell surrounding the CH_3I dopant as the perturber atoms become larger. The variation in the critical point effect, with krypton having a larger effect than argon and xenon, is caused by the strength of the perturber/CH_3I interactions in comparison to the perturber/perturber interactions, coupled with the differences in the ground-state and excited-state dopant/perturber interaction potentials. The CH_3I/Kr ground state potential well depth is close (i.e., 24 K) to the Kr/Kr potential well depth. This implies that the CH_3I/Kr interactions near the krypton critical point will be comparable to the Kr/Kr interactions, thereby leading to a large increase in the local perturber density near the critical point of the perturber, and a larger critical point effect. Similarly, the critical point effect decreases as one goes from krypton to argon to xenon because the difference in well depth for all intermolecular potentials increases.

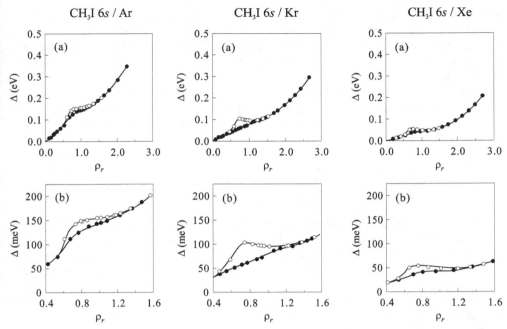

Figure 14. (a) The perturber induced shift $\Delta(\rho_P)$, as approximated by a moment analysis [i.e., eq. (21)], of the simulated primary transition for the CH_3I $6s$ Rydberg state as a function of the reduced perturber number density ρ_r for argon, krypton and xenon. ($\bullet$), simulations obtained at noncritical temperatures; ($\circ$), simulations near the critical isotherm. (b) An expanded view of $\Delta(\rho_P)$ near the perturber critical point. $\rho_c = 8.0 \times 10^{21}$ cm^{-3} for argon, $\rho_c = 6.6 \times 10^{21}$ cm^{-3} for krypton and $\rho_c = 5.0 \times 10^{21}$ cm^{-3} for xenon [12]. The solid lines provide a visual aid. See text for discussion.

5. Conclusion

In this work, the structure of low-n Rydberg states doped into supercritical fluids was investigated in several atomic perturbers. Both the experimental absorption spectra and full line shape simulations over the entire perturber density range at non-critical temperatures and along isotherms near perturber critical isotherms were presented for all dopant/perturber systems. These accurate line shape simulations allowed us to extract the perturber-induced energy shift $\Delta(\rho_P)$ from the simulated primary low-n Rydberg transitions. These shifts showed a striking critical point effect in all dopant/perturber systems. Our group also performed similar absorption measurements of atomic and molecular low-n Rydberg states in molecular perturbers [39, 40] with similar results. Because of the brevity of this Chapter, the details of these measurements cannot be presented here.

In all of the systems investigated [37–40], the dopant low-n Rydberg states are extremely sensitive to the nature of the perturbing fluid. When these states are doped into supercritical fluids, the surrounding perturbers interact with the central dopant causing shifts both in the dopant ground state energy and in the excited state energy. At low perturber number densities, the dopant/perturber interaction stabilizes the dopant ground state and the low-n Rydberg state. As the perturber density increases, perturber/dopant interactions lead to the formation of a perturber solvent shell around the dopant core, thereby inducing local perturber density inhomogeneities. This solvent shell begins to shield the optical electron from the cationic core. Therefore, the dense perturber fluid increases the dopant excitation energy, resulting in a blue shift of the abosrption band, which is observed experimentally. The local density of the first perturber solvent shell is almost proportional to the perturber bulk density at non-critical temperatures. However, near the critical isotherm and critical density of the perturber, the dopant/perturber interactions strengthen due to the increased perturber/perturber correlation length. This increased order yields a corresponding increase of the local density in the solvent shell that, in turn, leads to a stronger shielding of the optical electron from the cationic core. Thus, increased blue shifts of the low-n absorption bands are observed in all dopant/perturber systems near the critical point of the perturber. The area of this critical effect is demarcated by the turning points that bound the saddle point in the thermodynamic phase diagram of the critical isotherm.

For fluids with similar compressibilities, the structures of low-n dopant Rydberg states in the perturbing fluid show systematic behaviors. At non-critical temperatures, $\Delta(\rho_P)$ is determined by the polarizability and size of the perturbing fluid. The larger the polarizability and, therefore, the larger the size, the smaller the perturber-induced energy shift of the dopant absorption bands. This is caused by the number of atoms that can exist between the optical electron and the dopant cationic core, coupled with the strength of the shielding. The large overall energy shift observed in the dopant low-n Rydberg states perturbed by CF_4 [39, 40], on the other hand, was caused by the larger compressibility of CF_4 in comparison to the other gases in this study [37, 38, 40]. This larger compressibility implies that CF_4 is closer together on average at high perturber number densities than are the other perturbers studied, which increases the local density of CF_4 and, therefore, increases the blue shift in this perturber.

The critical point effect, on the other hand, is dominated by the similarity of the perturber/perturber interaction with the dopant/perturber ground state and dopant/perturber excited state interactions, coupled with the overall local density of the system. In krypton, the well depth of the ground state perturber/perturber intermolecular potential and the dopant/perturber intermolecular potential shows greater similarity in comparison to that in Ar and Xe. Moreover, the excited state CH_3I/Kr interaction is slightly stronger than the ground state Kr/Kr interaction. These facts dictate that the largest critical point effect for CH_3I in atomic perturbers is in Kr. Similarly, the largest overall critical effect was observed in CH_3I/CH_4 [39, 40]. This large critical effect is caused by both the ground state and excited state CH_3I/CH_4 interactions having strengths comparable to the CH_4/CH_4 interaction. Although the excited state CH_3I/CF_4 interactions are comparable in strength to the CF_4/CF_4 interactions, the ground state CH_3I/CF_4 interactions are not close to those of CF_4/CF_4. Similarly, the Xe/CF_4 ground state interactions are comparable to the ground state CF_4/CF_4 interactions, but the excited state Xe/ground state CF_4 interactions are weaker. Moreover, the bulk critical density in CF_4 is small in comparison to the rest of the perturbers investigated here. This results in the CF_4 critical effect on $\Delta(\rho_P)$ being the smallest one observed [39, 40].

These data sets also allowed us to generate a consistent set of intermolecular potential parameters for various dopant/perturber systems, which are summarized in Appendix A. Several general trends in these parameters can be observed. For atomic perturbers, the steepness of the exponential-6 intermolecular potential (i.e., γ) used to model the dopant excited state/perturber intermolecular interaction decreases with increasing perturber size and polarizability. This trend is reversed in molecular perturbers, were the larger, more compressible CF_4 has a steeper repulsive component in comparison to CH_4. The excited vibrational states of CH_3I always have exponential-6 potentials with a smaller γ in comparison to the CH_3I adiabatic transition in the same perturbing gas. Moreover, the vibrational states always have an equilibrium collision radius that is identical or larger than the collision radius of the adiabatic transition. The excited state collision radii are always larger than the ground state collision radii, as one would expect. However, the interaction strength of the excited state (as gauged by the well depth) can be stronger or weaker than that for the ground state of the same system. These changing interactions are what dominate the variations observed in the critical effects for each of the dopant/perturber systems investigated here.

An understanding of the structure of low-n Rydberg states in supercritical fluids is an important tool in the investigation of solvation effects, since these studies can yield accurate dopant/perturber ground state and excited state intermolecular potentials. We conclude from the present work that the absorption line shapes can be adequately simulated within a simple semi-classical line shape analysis. However, this work focused on highly symmetric perturbers. Future studies should concern more asymmetric perturbers and polar perturbers. Such an extension will require changing the calculation techniques involved in determining the radial distribution functions as well as the type of Fourier transform used to simulate the line shape. Since the excited state is sensitive to the structure of the perturbing fluid, we anticipate that multi-site intermolecular potentials and angular dependent intermolecular potentials will be needed as the perturber complexity increases, in order to model the full line shape accurately.

Acknowledgements

All experimental measurements were made at the University of Wisconsin Synchrotron Radiation Center (NSF DMR-0537588), with support from the Petroleum Research Fund (PRF#45728-B6), the Professional Staff Congress - City University of New York, the Louisiana Board of Regents Support Fund (LEQSF(2006-09)-RD-A-33), and the National Science Foundation (NSF CHE-0956719).

Author details

Luxi Li and Xianbo Shi
Brookhaven National Laboratory, Upton, NY, USA

Cherice M. Evans
Department of Chemistry, Queens College – CUNY and the Graduate Center – CUNY, New York, NY, USA

Gary L. Findley
Chemistry Department, University of Louisiana at Monroe, Monroe, LA, USA

Appendix A. Intermolecular potential parameters

Below is a tablulated list of the intermolecular potential parameters used to simulate the absorption line shapes in the various dopant/perturber systems presented or summarized in this work [40].

	ε/k_B (K)	r_e (Å)	γ	Ref.
Ar/Ar	119.5	3.826	–	[37, 38]
Kr/Kr	172.7	4.031	–	[38]
Xe/Xe	229.0	4.552	–	[38]
$CH_4/CH_4{}^a$	141.5	3.704	–	[39]
CF_4/CF_4	181.02	4.708	–	[39]
Xe/Ar	200.0	4.265	–	[37]
XeAr/Ar	195.0	4.310	–	[37]
Xe/CF_4	199.3	4.629	–	[39]
CH_3I/Ar	162.2	4.572	–	[38]
CH_3I/Kr	196.7	4.676	–	[38]
CH_3I/Xe	297.5	4.896	–	[38]
CH_3I/CH_4	195.8	4.243	–	[39]
CH_3I/CF_4	256.0	5.016	–	[39]
Xe $6s$/Ar	300.0	5.20	16.00	[37]
Xe $6s/CF_4$	135.0	6.55	12.25	[39]
CH_3I $6s$/Ar	110.0	6.30	12.75	[38]
CH_3I $6s$/Kr	245.0	6.20	11.30	[38]
CH_3I $6s$/Xe	400.0	6.39	10.25	[38]
CH_3I $6s/CH_4$	145.0	6.55	10.10	[39]
CH_3I $6s/CF_4$	185.0	6.84	12.10	[39]
Xe $6s'$/Ar	400.0	4.98	16.00	[37]
CH_3I $6s'$/Ar	110.0	6.30	12.75	[38]
CH_3I $6s'$/Kr	245.0	6.20	11.30	[38]
CH_3I $6s'$/Xe	400.0	6.29	10.25	[38]
CH_3I $6s'/CH_4$	145.0	6.55	10.10	[39]
CH_3I $6s'/CF_4$	185.0	6.84	12.10	[39]
Xe$(6s)$Ar/Ar	250.0	5.25	16.00	[37]
CH_3I $6s$ ν_2/Ar	150.0	6.30	12.15	[38]
CH_3I $6s$ ν_2/Kr	225.0	6.30	10.75	[38]
CH_3I $6s$ ν_2/Xe	360.0	6.50	9.50	[38]
CH_3I $6s$ ν_2/CH_4	105.0	6.65	9.95	[39]
CH_3I $6s$ ν_2/CF_4	135.0	6.84	11.90	[39]
CH_3I $6s'$ ν_2/Ar	150.0	6.30	12.15	[38]
CH_3I $6s'$ ν_2/Kr	225.0	6.30	10.75	[38]
CH_3I $6s'$ ν_2/Xe	360.0	6.35	9.50	[38]
CH_3I $6s'$ ν_2/CH_4	105.0	6.65	9.95	[39]
CH_3I $6s'$ ν_2/CF_4	135.0	6.84	11.90	[39]

^{a} Two-Yukawa potential with $\kappa_0 = 8.50$ Å, $z_1 = 0.90$ Å^{-1}, and $z_2 = 4.25$ Å^{-1}.

6. References

[1] M. B. Robin. *Higher Excited States of Polyatomic Molecules Vol. I – Vol. III.* Academic Press, New York, 1974, 1975, 1985. and references therein.

[2] R. Reininger, U. Asaf, I. T. Steinberger, and S. Basak. Relationship between the energy v_0 of the quasi-free-electron and its mobility in fluid argon, krypton, and xenon. *Physical Review B*, 28(8):4426–4432, 1983.

[3] R. Reininger, U. Asaf, and I. T. Steinberger. The density dependence of the quasi-free electron state in fluid xenon and krypton. *Chemical Physics Letters*, 90(4):287–290, 1982.

[4] A. O. Allen and W. F. Schmidt. Determination of the energy level v_0 of electrons in liquid argon over a range of densities. *Zeitschrift für Naturforschung A*, 37(4):316–318, 1982.

[5] W. von Zdrojewski, J. G. Rabe, and W. F. Schmidt. Photoelectric determination of v_0-values in solid rare gases. *Zeitschrift fur Naturforschung A*, 35(7):672–674, 1980.

[6] B. Halpern, J. Lekner, S. A. Rice, and R. Gomer. Drift velocity and energy of electrons in liquid argon. *Physical Review*, 156(2):351–352, 1967.

[7] W. Tauchert, H. Jungblut, and W. F. Schmidt. Photoelectric determination of v_0 values and electron ranges in some cryogenic liquids. *Canadian Journal of Chemistry*, 55(11):1860–1866, 1977.

[8] J. R. Broomall, W. D. Johnson, and D. G. Onn. Density dependence of the electron surface barrier for fluid helium-3 and helium-4. *Physical Review B*, 14(7):2819–2825, 1976.

[9] J. Jortner and A. Gaathon. Effects of phase density on ionization processes and electron localization in fluids. *Canadian Journal of Chemistry*, 55(11):1801–1819, 1977.

[10] N. Schwenter, E. E. Koch, and J. Jortner. *Electronic Excitations in Condensed Rare Gases.* Springer-Verlag, Berlin, 1985.

[11] U. Asaf, R. Reininger, and I. T. Steinberger. The energy v_0 of the quasi-free electron in gaseous, liquid, and solid methane. *Chemical Physics Letters*, 100:363–366, 1983.

[12] C. M. Evans and G. L. Findley. Energy of the quasifree electron in argon and krypton. *Physical Review A*, 72:022717, 2005.

[13] C. M. Evans and G. L. Findley. Energy of the quasi-free electron in supercritical argon near the critical point. *Chemical Physics Letters*, 410:242–246, 2005.

[14] C. M. Evans and G. L. Findley. Field ionization of c2h5i in supercritical argon near the critical point. *Journal of Physics B: Atomic, Molecular and Optical Physics*, 38:L269–L275, 2005.

[15] Luxi Li, C. M. Evans, and G. L. Findley. Energy of the quasi-free electron in supercritical krypton near the critical point. *Journal of Physical Chemistry A*, 109:10683–10688, 2005.

[16] Xianbo Shi, Luxi Li, C. M. Evans, and G. L. Findley. Energy of the quasi-free electron in xenon. *Chemical Physics Letters*, 432:62–67, 2006.

[17] Xianbo Shi, Luxi Li, C. M. Evans, and G. L. Findley. Energy of the quasi-free electron in argon, krypton and xenon. *Nuclear Instruments and Methods in Physics Research A*, 582:270–273, 2007.

[18] Xianbo Shi, Luxi Li, G. M. Moriarty, C. M. Evans, and G. L. Findley. Energy of the quasi-free electron in low density ar and kr: extension of the local wigner-seitz model. *Chemical Physics Letters*, 454:12–16, 2008.

[19] Xianbo Shi, Luxi Li, G. L. Findley, and C. M. Evans. Energy of the excess electron in methane and ethane near the critical point. *Chemical Physics Letters*, 481:183–189, 2009.

[20] G. D. Mahan. Satellite bands in alkali-atom spectra. *Physical Review A*, 6:1273–1279, 1972.

[21] M. Lax. The franck-condon principle and its application to crystals. *Journal of Chemical Physics*, 20:1752–1760, 1952.

[22] R. Granier, M. C. Castex, J. Granier, and J. Romand. Perturbation of the xenon 1469 a. resonance line by various rare gases and hydrogen. *Comptes Rendus de l'Académie des Sciences B*, 264:778, 1967.

[23] M. C. Castex, R. Granier, and J. Romand. Perturbation of the 1236-a. resonance line of krypton and the 1295-a. resonance line of xenon by various rare gases. *Comptes Rendus de l'Académie des Sciences B*, 268:552, 1969.

[24] M. C. Castex. Absorption spectra of xenon-rare gas mixtures in the far uv region (1150-1500 å): high resolution analysis and first quantitative absorption measurements. *Journal of Chemical Physics*, 66:3854–3865, 1977.

[25] I. Messing, B. Raz, and J. Jortner. Medium perturbations of atomic extravalence excitations. *Journal of Chemical Physics*, 66:2239–2251, 1977.

[26] I. Messing, B. Raz, and J. Jortner. Solvent perturbations of extravalence excitations of atomic xenon by rare gases at high pressures. *Journal of Chemical Physics*, 66:4577–4586, 1977.

[27] I. Messing, B. Raz, and J. Jortner. Perturbations of molecular extravalence excitations by rare-gas fluids. *Chemical Physics*, 25:55–74, 1977.

[28] I. Messing, B. Raz, and J. Jortner. Medium effects on the vibrational structure of some molecular rydberg excitations. *Chemical Physics*, 23:351–355, 1977.

[29] A. M. Halpern. Iterative fourier reconvolution spectroscopy: van der waals broadening of rydberg transitions; the $\sim$b $\leftarrow$ $\sim$x (5p, 6s) transition of methyl iodide. *Journal of Physical Chemistry*, 96:2448–2455, 1992.

[30] T. Kalbfleisch, R. Fan, J. Roebber, P. Moore, E. Jacobson, and L.D. Ziegler. A molecular dynamics study of electronic absorption line broadening in high-pressure nonpolar gases. *Journal of Chemical Physics*, 103:7673–7684, 1995.

[31] R. Fan, T. Kalbfleisch, and L. D. Ziegler. A molecular dynamics analysis of resonance emission: optical dephasing and inhomogeneous broadening of ch_3i in ch_4 and ar. *Journal of Chemical Physics*, 104:3886–3897, 1996.

[32] T. S. Kalbfleisch, L. D. Ziegler, and T. Keyes. An instantaneous normal mode analysis of solvation: methyl iodide in high pressure gases. *Journal of Chemical Physics*, 105:7034–7046, 1996.

[33] S. A. Egorov, M. D. Stephens, and J. L. Skinner. Absorption line shapes and solvation dynamics of ch_3i in supercritical ar. *Journal of Chemical Physics*, 107:10485–10491, 1997.

[34] E. Morikawa, A. M. Köhler, R. Reininger, V. Saile, and P. Laporte. Medium effects on valence and low-n rydberg states: No in argon and krypton. *Journal of Chemical Physics*, 89:2729–2737, 1988.

[35] P. Larrégaray, A. Cavina, and M. Chergui. Ultrafast solvent response upon a change of the solute size in non-polar supercritical fluids. *Chemical Physics*, 308:13–25, 2005.

[36] C. N. Tiftickjian and S. A. Egorov. Absorption and emission lineshapes and solvation dynamics of no in supercritical ar. *Journal of Chemical Physics*, 128:114501, 2008.

[37] Luxi Li, Xianbo Shi, C. M. Evans, and G. L. Findley. Xenon low-n rydberg states in supercritical argon near the critical point. *Chemical Physics Letters*, 461:207–210, 2008.

[38] Luxi Li, Xianbo Shi, C. M. Evans, and G. L. Findley. Ch$_3$i low-n rydberg states in supercritical atomic fluids near the critical point. *Chemical Physics*, 360:7–12, 2009.

[39] Luxi Li, Xianbo Shi, G. L. Findley, and C. M. Evans. Dopant low-n rydberg states in cf_4 and ch_4 near the critical point. *Chemical Physics Letters*, 482:50–55, 2009.

[40] Luxi Li. *Atomic and Molecular Low-n Rydberg States in Near Critical Point Fluids*. PhD thesis, The Graduate Center of the City University of New York, New York, NY, 2009.

[41] S. C. Tucker. Solvent density inhomogeneities in supercritical fluids. *Chemical Reviews*, 99:391–418, 1999.

[42] P. Attard. Spherically inhomogeneous fluids. i. percus-yevick hard spheres: osmotic coefficients and triplet correlations. *Journal of Chemical Physics*, 91:3072–3082, 1989.

[43] P. Attard. Spherically inhomogeneous fluids. ii. hard-sphere solute in a hard-sphere solvent. *Journal of Chemical Physics*, 91:3083–3089, 1989.

[44] J. Zhang, D. P. Roek, J. E. Chateauneuf, and J. F. Brennecke. A steady-state and time-resolved fluorescence study of quenching reactions of anthracene and 1,2-benzanthracene by carbon tetrabromide and bromoethane in supercritical carbon dioxide. *Journal of the American Chemical Society*, 119:9980–9991, 1997.

[45] Y. P. Sun, M. A. Fox, and K. P. Johnston. Spectroscopic studies of p-(n,n-dimethylamino)benzonitrile and ethyl p-(n,n-dimethylamino)benzoate in supercritical trifluoromethane, carbon dioxide, and ethane. *Journal of the American Chemical Society*, 114:1187–1194, 1992.

[46] R. S. Urdahl, D. J. Myers, K. D. Rector, P. H. Davis, B. J. Cherayil, and M. D. Fayer. Vibrational lifetimes and vibrational line positions in polyatomic supercritical fluids near the critical point. *Journal of Chemical Physics*, 107:3747–3757, 1997.

[47] R. S. Urdahl, K. D. Rector, D. J. Myers, P. H. Davis, and M. D. Fayer. Vibrational relaxation of a polyatomic solute in a polyatomic supercritical fluid near the critical point. *Journal of Chemical Physics*, 105:8973–8976, 1996.

[48] R. P. Futrelle. Unified theory of spectral line broadening in gases. *Physical Review A*, 5:2162–2182, 1972.

[49] J. L. Lebowitz and J. K. Percus. Statistical thermodynamics of nonuniform fluids. *Journal of Mathematical Physics*, 4:116–123, 1963.

[50] J. L. Lebowitz and J. K. Percus. Asymptotic behavior of radial distribution function. *Journal of Mathematical Physics*, 4:248–254, 1963.

[51] J. L. Lebowitz and J. K. Percus. Integral equations and inequalitites in theory of fluids. *Journal of Mathematical Physics*, 4:1495–1506, 1963.

[52] M. Born and H. S. Green. A general kinetic theory of liquids. i. the molecular distribution functions. *Proceedings of the Royal Society of London Series A*, 188:10–18, 1946.

[53] R. Kubo and Y. Toyozawa. Application of the method of generating function to radiative and non-radiative transitions of a trapped electron in a crystal. *Progress of Theoretical Physics*, 13:160–182, 1955.

[54] H. C. Jacobson. Moment analysis of atomic spectral lines. *Physical Review A*, 4:1363–1368, 1971.

[55] T. R. Strobridge. The thermodynamic properties of nitrogen from 64 to 300 °k between 0.1 and 200 atmospheres. *NBS Technical Note*, 129, 1962.

[56] A. L. Gosman, R. D. McCarty, and J. G. Hust. Thermodynamics properties of argon from the triple point to 300 k at pressures to 1000 atmostpheres. *NBS Technical Note*, 27, 1969.

[57] W. B. Streett and L. A. K. Staveley. Experimental study of the equation of state of liquid krypton. *Journal of Chemical Physics*, 55:2495–2506, 1971.

[58] W. B. Streett, L. S. Sagan, and L. A. K. Staveley. Experimental study of the equation of state of liquid xenon. *Journal of Chemical Thermodynamics*, 5:633–650, 1973.

[59] E. W. Grundke, D. Henderson, and R. D. Murphy. Evaluation of the percus-yevick theory for mixtures of simple liquids. *Canadian Journal of Physics*, 51:1216–1226, 1973.

[60] W. H. Press, S. A. Teukolsky, W. T. Vetterling, and B. P. Flannery. *Numerical Recipes in FORTRAN: The Art of Scientific Computing*. Cambridge University Press, New York, 1992.

[61] Xianbo Shi. *Energy of the Quasi-free Electron in Atomic and Molecular Fluids*. PhD thesis, The Graduate Center of the City University of New York, New York, NY, 2010.

Chapter 7

Injection and Optical Spectroscopy of Localized States in II-VI Semiconductor Films

Denys Kurbatov, Anatoliy Opanasyuk and Halyna Khlyap

Additional information is available at the end of the chapter

1. Introduction

Novel achievements of nano- and microelectronics are closely connected with working-out of new semiconductor materials. Among them the compounds II-VI (where A = Cd, Zn, Hg and B = O, S, Se, Te) are of special interest. Due to unique physical properties these materials are applicable for design of optical, acoustical, electronic, optoelectronic and nuclear and other devices [1-3]. First of all the chalcogenide compounds are direct gap semiconductors where the gap value belongs to interval from 0.01 eV (mercury chalcogenides) up to 3.72 eV (ZnS with zinc blende crystalline structure) As potential active elements of optoelectronics they allow overlapping the spectral range from 0.3 μm to tens μm if using them as photodetectors and sources of coherent and incoherent light. The crystalline structure of II-VI compounds is cubic and hexagonal without the center of symmetry is a good condition for appearing strong piezoeffect. Crystals with the hexagonal structure have also pyroelectric properties. This feature may be used for designing acoustoelectronic devices, amplifiers, active delay lines, detectors, tensile sensors, etc. [1-2]. Large density of some semiconductors (CdTe, ZnTe, CdSe) makes them suitable for detectors of hard radiation and α–particles flow [4-5]. The mutual solubility is also important property of these materials. Their solid solutions give possibility to design new structures with in-advance defined gap value and parameters of the crystalline lattice, transmission region, etc. [6].

Poly- and monocrystalline films of II-VI semiconductors are belonging to leaders in field of scientific interest during the last decades because of possibility of constructing numerous devices of opto-, photo-, acoustoelectronics and solar cells and modules [2-5]. However, there are also challenges the scientists are faced due to structural peculiarities of thin chalcogenide layers which are determining their electro-physical and optical characteristics. The basic requirements for structure of thin films suitable for manufacturing various microelectronic devices are as follows: preparing stoichiometric single phase monocrystalline layers or columnar strongly textured polycrystalline layers with low concentration of stacking faults (SF), dislocations, twins with governed ensemble of point defects (PD) [7-8]. However, an enormous number of publications points out the following features of these films: tend to departure of stioichiometric composition, co-existing two polymorph modifications (sphalerite and wurtzite), lamination morphology of crystalline grains (alternation of cubic and hexagonal phases), high concentration of twins and SF, high level of micro- and macrostresses, tend to formation of anomalous axial structures, etc. [2-3, 9]. Presence of different defects which are recombination centers and deep traps does not improve electro-physical and optical characteristics of chalcogenide layers. It restricts the application of the binary films as detector material, basic layers of solar energy photoconvertors, etc.

Thus, the problem of manufacturing chalcogenide films with controllable properties for device construction is basically closed to the governing of their defect structure investigated in detail. We will limit our work to the description of results from the examination of parameters of localized states (LS) in polycrystalline films CdTe, ZnS, ZnTe by the methods of injection and optical spectroscopy.

1.1. Defect classification in layers of II-VI compounds

Defects' presence (in the most cases the defects of the structure are charged) is an important factor affecting structure-depended properties of II-VI compounds [3, 5, 10]. Defects of the crystalline structure are commonly PD, 1-, 2-, and 3-dimensional ones [11-12]. Vacancies (V_A, V_B), interstitial atoms (A_i, B_i), antistructural defects (A_B, B_A), impurity atoms located in the lattice sites (C_A, C_B) and in the intersites (C_i) of the lattice are defects of the first type. However, the antistructural defects are not typical for wide gap materials (except CdTe) and they appear mostly after ionizing irradiation [13-14]. The PD in chalcogenides can be one- or two-charged. Each charged native defect forms LS in the gap of the semiconductor, the energy of the LS is ΔE_i either near the conduction band (the defect is a donor) or near the valence band (then the defect is an acceptor) as well as LS formed in energy depth are appearing as traps for charge carriers or recombination centers [15-16]. Corresponding levels in the gap are called shallow or deep LS. If the extensive defects are minimized the structure-depending properties of chalcogenides are principally defined by their PD. The effect of traps and recombination centers on electrical characteristics of the semiconductor materials is considered in [16]. We have to note that despite a numerous amount of publications about PD in Zn-Cd chalcogenides there is no unified theory concerning the nature of electrically active defects for the range of chalcogenide vapor high pressures as well as for the interval of high vapor pressure of chalcogen [13-14, 17-18].

Screw and edge dislocations are defects of second type they can be localized in the bulk of the crystalites or they form low-angled boundaries of regions of coherent scattering (RCS). Grain boundaries, twins and surfaces of crystals and films are defects of the third type. Pores and precipitates are of the 4th type of defects. All defects listed above are sufficiently influencing on physical characteristics of the real crystals and films of II-VI compounds due to formation of LS (along with the PD) in the gap of different energy levels [17-20].

2. Using injection spectroscopy for determining parameters of localized states in II-VI compounds

2.1. Theoretical background of the injection spectroscopy method

The LS in the gap of the semiconductor make important contribution to the function of the device manufactured from the material solar cells, phodetectors, γ-ray detectors and others), for example, carriers' lifetime, length of the free path, etc., thus making their examination one of them most important problems of the semiconductor material science [3-5, 8, 13, 14, 18].

There are various methods for investigation of the energy position (E_t), concentration (N_t) and the energy distribution of the LS [21-23]. However, their applicability is restricted by the resistance of the semiconductors, and almost all techniques are suitable for low-resistant semiconductors. At the investigation of the wide gap materials II-VI the analysis of current-voltage characteristics (CVC) at the mode of the space-charge limited current (SCLC) had appeared as a reliable tool [24-25]. The comparison of experimental and theoretical CVCs is carried out for different trap distribution: discrete, uniform, exponential, double-exponential, Gaussian and others [26-36]. This method is a so-called direct task of the experiment and gives undesirable errors due to in-advance defined type of the LS distribution model used in further working-out of the experimental data. The information obtained is sometimes unreliable and incorrect.

Authors [37-40] have proposed novel method allowing reconstructing the LS energy distribution immediately from the SCLC CVC without the pre-defined model (the reverse task), for example, for organic materials with energetically wide LS distributions [41-42]. However, the expressions presented in [37-40], as shown by our studies [43-45], are not suitable for analysis of experimental data for mono- and polycrystalline samples with energetically narrow trap distributions. So, we use the principle [37-40] and obtain reliable and practically applicable expressions for working-out of the real experiments performed for traditional II-VI compounds.

Solving the Poisson equation and the continuity equation produces SCLC CVC for rectangular semiconductor samples with traps and deposited metallic contacts, where the source contact (cathode) provides charge carriers' injection in the material [24-25]:

$$j = e\mu E(x)n_f(x), \tag{1}$$

$$\frac{dE(x)}{dx} = \frac{e\left[\left(n_f(x) - n_{f0}\right) + \sum_j \left(n_{t_j}(x) - n_{t_{j0}}\right)\right]}{\varepsilon\varepsilon_0} = \frac{en_s(x)}{\varepsilon\varepsilon_0}, \tag{2}$$

where j current density passes through the sample;
e electron charge;
μ drift carrier mobility;
ε_0 dielectric constant;
ε permittivity of the material

$E(x)$ is an external electric field changing by the depth of the sample; this field injects free carriers from the source contact (cathode) ($x=0$) to the anode collecting the carriers ($x=d$);

$n_f(x)$ is the free carriers' concentration at the injection;

n_{f0} is the equilibrium free carriers concentration;

$n_{t_j}(x)$ is the concentration of carriers confined by the traps of the j-group with the energy level E_{t_j};

n_{tj0} is the equilibrium carriers concentration trapped by the centers of the j-group;

$n_s(x)$ is a total concentration of the injected carriers.

The set of equations (1), (2) is commonly being solved with a boundary condition $E(0)=0$. The set is soluble if the function from n_f and n_t is known. We assume that all LS in the material are at thermodynamic equilibrium with corresponding free bands, then their filling-in by the free carriers is defined by the position of the Fermi quasi-level E_F. Using the Boltzmann statistics for free carriers and the Fermi – Dirac statistics for the localized carriers we can write [39-40]:

$$n_f(x) = N_{C(V)} \exp\left(\frac{E_{C(V)}(x) - E_F(x)}{kT}\right),$$ (3)

$$n_t(E,x) = \frac{h(E,x)}{1 + g\exp\dfrac{E_t(x) - E_F(x)}{kT}},$$ (4)

where $N_{C(V)}$ are states density in conduction band (valence band);
$E_{C(V)}$ is energy of conduction band bottom (valence band top);
k is Boltzmann constant;
T is the temperature of measurements;
$E_F(x)$ is the Fermi quasi-level at injection;

g is a factor of the spin degeneration of the LS which depends on its charge state having the following values: –1/2, 1 or 2 (typically $g = 1$) [15, 39-40].

The zero reference of the trap energy level in the gap of the material will be defined relatively to the conduction band or valence band depending on the type (n or p) of the examined material: $E_{C(V)}=0$.

The set of equations (1)–(2) can also be reduced to integral relations. Detailed determination of these ratios presented in [37].

$$\frac{1}{j} = \frac{1}{e\mu d}\frac{\varepsilon\varepsilon_0}{e}\int_{n_{fc}}^{n_{fa}}\frac{dn_f}{n_f^2[(n_f - n_{f_0}) + \sum_j (n_{t_j} - n_{t_{j_0}})]} \equiv y, \tag{5}$$

$$\frac{U}{j^2} = \frac{\varepsilon\varepsilon_0}{e(e\mu)^2}\int_{n_{fc}}^{n_{fa}}\frac{dn_f}{n_f^3[(n_f - n_{f_0}) + \sum_j (n_{t_j} - n_{t_{j_0}})]} \equiv z, \tag{6}$$

where j, U are current density and voltage applied to the sample;
d is the sample thickness;
n_{fc}, n_{fa} are free carriers' concentration in cathode and anode, respectively.

Equations (5) and (6) determine SCLC CVCs in parametric form for an arbitrary distribution of LS in the gap of the material.

At thermodynamic equilibrium the total concentration (n_{s_0}), the carriers concentration for those localized on the traps (n_{t_0}), and the free carriers' concentration in the semiconductor (n_{f_0}) are in the function written as follows: $n_{s_0} = n_{t_0} + n_{f_0}$, де $n_{f_0} = N_{C(V)}\exp\left(-\dfrac{E_{C(V)} - E_{F_0}}{kT}\right)$ in case when $E_c - E_{F0} \geq 3kT$ ($3kT$= 0,078 eV at the room temperature; E_{F0} is the equilibrium Fermi level. It must be emphasized that this charge limits the current flow through the sample and determines the form of the SCLC CVC.

The carriers' injection from the source contact leads to appearance of the space charge ρ in the sample, formed by the free carriers and charge carriers localized in the traps, $\rho = en_i = e(n_s - n_{s_0}) = e[(n_t + n_f) - (n_{t_0} + n_{f_0})]$, where n_i is the concentration of injected carriers.

Under SCLC mode the concentration of injected carriers is considerably larger than their equilibrium concentration in the material and, at the same time, it is sufficiently lower than the total concentration of the trap centers ($n_{f_0} \ll n_i \ll N_t$) [24-25]. Thus, in further description we will neglect the second term in the expression written above (except some special cases). Then we have $\rho = en_s(x) \sim e\sum_j n_{t_j}(x)$.

Using (5) and (6) we find the first and second derivatives of z from y:

$$z' = \frac{dz}{dy} = \frac{d(U/j^2)}{d(1/j)} = \frac{d}{e\mu n_{fa}}, \tag{7}$$

$$z'' = \frac{d^2 z}{dy^2} = \frac{d}{d(1/j)} \frac{d(U/j^2)}{d(1/j)} = \left| \frac{\rho_a d^2}{\varepsilon \varepsilon_0} \right|. \tag{8}$$

As the SCLC CVC are commonly represented in double-log scale [24-25], equations (7), (8) are rewritten with using derivatives: $\eta = \frac{d(\ln j)}{d(\ln U)}$, $\eta' = \frac{d^2(\ln j)}{d(\ln U)^2}$, $\eta'' = \frac{d^3(\ln j)}{d(\ln U)^3}$.

Then we have

$$n_{fa} = \frac{\eta}{2\eta - 1} \frac{jd}{e\mu U} = \frac{1}{\alpha} \frac{jd}{e\mu U}, \tag{9}$$

$$\frac{\rho_a}{e} = \frac{2\eta - 1}{\eta} \frac{\eta - 1}{\eta} \left[1 - \frac{\eta'}{\eta(2\eta - 1)(\eta - 1)} \right] \frac{\varepsilon \varepsilon_0 U}{ed^2} = \alpha \beta \frac{\varepsilon \varepsilon_0 U}{ed^2}, \tag{10}$$

where $\alpha = \frac{2\eta - 1}{\eta}$, $\beta = \frac{\eta - 1}{\eta} \left[1 - \frac{\eta'}{\eta(2\eta - 1)(\eta - 1)} \right] = \frac{\eta - 1}{\eta}(1 + B)$.

Further we will neglect the index a.

As a result, the Poisson equation and the continuity equation give fundamental expressions for a dependence of the free carrier concentration in the sample n_f (the Fermi quasi-level energy) and space charge density at the anode ρ on the voltage U and the density of the current j flowing through the structure metal-semiconductor-metal (MSM).

Now let us consider the practical application of expressions (7) and (8) or (9) and (10) for reconstructing the trap distribution in the gap of the investigated material. We would restrict with the electron injection into n-semiconductor.

If the external voltage changes the carries are injected from the contact into semiconductor; at the same time, the Fermi quasi-level begins to move between the LS distributed in the gap from the start energy E_{F0} up to conduction band. This displacement ΔE_F leads to filling-in of the traps with the charge carriers and, consequently, to the change of the conductivity of the structure. Correspondingly, under intercepting the Fermi quasi-level and the monoenergetical LS the CVC demonstrates a peculiarity of the current [24-25]. As the voltage and current density are in the function of the LS concentration with in-advanced energy position and the Fermi quasi-level value we obtain a possibility to scan the energy distributions. This relationship is a physical base of the injection spectroscopy method (IS).

Increase of the charge carriers dn_s in the material at a low change of the Fermi level position is to be found from the expression:

$$\frac{1}{e} \frac{d\rho}{dE_F} = \frac{dn_i}{dE_F} \approx \frac{dn_s}{dE_F} \approx \frac{dn_t}{dE_F}. \tag{11}$$

The carrier concentration on deep states can be found from the Fermi-Dirac statistics

$$n_s = \int_{E_1}^{E_2} n_s(E)dE = \int_{E_1}^{E_2} h(E)f(E - E_F)dE + n_f, \tag{12}$$

where $dn_s(E)/dE$ is a function describing the energy distribution of trapped carriers; $h(E)=dN_t/dE$ is a function standing for the energy trap distribution; E_1, E_2 are energies of start and end points for the LS distribution in the gap of the material.

It is assumed that the space trap distribution in the semiconductor is homogeneous by the sample thickness then $h(E,x) = h(E)$.

After substitution of (12) in (11) we obtain a working expression for the functions $d\rho/dE_F$ and $h(E)$

$$\frac{1}{e}\frac{d\rho}{dE_F} \approx \frac{dn_s}{dE_F} = \frac{d}{dE_F}\int_E n_s(E)dE = \int_E h(E)\frac{d(f(E - E_F))}{d(E - E_F)} + \frac{n_f}{kT} \tag{13}$$

Thus, at arbitrary temperatures of the experiment the task of reconstructing LS distributions reduces to finding function $h(E)$ from the convolution (12) or (13) using known functions $n_s(E_F)$ or $d\rho/dE_F$. The expression (12) is the most preferable [39-40]. In general case the solution is complex and it means determining the function $h(E)$ from the convolution (12) or (13) if one of the functions n_s or dn_s/dE_F is known [43-45]. We have solved this task according the Tikhonov regularization method [46]. If the experiment is carried at low temperatures (liquid nitrogen) the problem is simplified while the Fermi-Dirac function in (13) may be replaced with the Heavyside function and, neglecting n_f, we obtain

$$\frac{1}{e}\frac{d\rho}{dE_F} \approx \frac{dN_t}{dE_F} \approx h(E). \tag{14}$$

This equation shows that the function $1/e \; d\rho/dE_F$ - E_F at low-temperature approximation immediately produces the trap distribution in the gap of the semiconductor. Using (7) and (8), we transform the expression (14) for practical working-out of the experimental SCLC CVC. As the free carrier concentration and the space charge density are to be written as follows:

$$n_f = \frac{d}{e\mu}\frac{j}{2U - U'j}, \tag{15}$$

$$\frac{\rho}{e} = (U''j^2 - 2U'j + 2U)\frac{\varepsilon\varepsilon_0}{ed^2}, \tag{16}$$

the expression (14) will be

$$h(E) \approx \frac{1}{e}\frac{d\rho}{dE_F} = \frac{1}{kT}\frac{\varepsilon\varepsilon_0}{ed^2}\frac{U'''j^3(2U - U'j)}{(U''j^2 - 2U'j + 2U)}. \tag{17}$$

Using derivatives η, η', η'' this expression is easily rewritten:

$$h(E) \approx \frac{1}{e}\frac{d\rho}{dE_F} = \frac{1}{kT}\frac{\varepsilon\varepsilon_0 U}{ed^2}\frac{2\eta-1}{\eta^2}\left[1+\frac{(3\eta-3)\eta\eta'-\eta\eta''+3\eta'^2}{\eta^2\left((2\eta-1)(\eta-1)-\eta'/2\right)}\right]. \tag{18}$$

The expression (18) is also can be written with the first derivative (η) only. Denote

$$C = \frac{(3\eta-3)\eta\eta'-\eta\eta''+3\eta'^2}{\eta^2\left[(2\eta-1)(\eta-1)-\eta'/2\right]} = (2-3\eta)B + \frac{d\ln(1+B)}{d\ln U} = \frac{(2\eta-1)B+(3\eta-2)B^2+\dfrac{d\left[\ln(1+B)\right]}{d\ln U}}{1+(\eta-1)B},$$

where $B = -\dfrac{1}{\eta(2\eta-1)(\eta-1)}\dfrac{d\eta}{d\ln U}$.

We obtained an expression used by authors [39-40] for analysis of energetically wide LS distributions in organic semiconductors.

$$h(E) \approx \frac{1}{e}\frac{d\rho}{dE_F} = \frac{1}{kT}\frac{\varepsilon\varepsilon_0 U}{ed^2}\frac{2\eta-1}{\eta^2}(1+C) = \frac{\alpha\beta}{kT}\frac{\varepsilon\varepsilon_0 U}{ed^2}\frac{1}{\eta-1}\left(\frac{1+C}{1+B}\right). \tag{19}$$

To make these expressions suitable for the working-out of SCLC CVC for the semiconductors with energetically narrow trap distributions we write them with reverse derivatives $\gamma = \dfrac{d(\ln U)}{d(\ln j)}$, $\gamma' = \dfrac{d^2(\ln U)}{d(\ln j)^2}$, $\gamma'' = \dfrac{d^3(\ln U)}{d(\ln j)^3}$.

As a result:

$$h(E) \approx \frac{1}{e}\frac{d\rho}{dE_F} = \frac{1}{kT}\frac{\varepsilon\varepsilon_0 U}{ed^2}\left[\frac{(2\gamma-3)\gamma'+\gamma''}{(2-\gamma)(1-\gamma)+\gamma'}+\gamma\right](2-\gamma). \tag{20}$$

Solving the set of equations (3) and (7) gives energetical scale under re-building deep trap distributions. Using various derivatives we obtain

$$E_F = kT\ln\frac{e\mu N_{C(V)}}{d} + kT\ln\frac{2U-U'j}{j} = kT\ln\frac{e\mu N_{C(V)}}{d} + kT\ln\frac{j}{U} + kT\ln\frac{\eta}{2\eta-1} =$$

$$= kT\ln\frac{e\mu N_{C(V)}}{d} + kT\ln\frac{j}{U} + kT\ln\frac{1}{2-\gamma}. \tag{21}$$

Using sets of equations (17) - (18) or (20) - (21) allows to find a function describing the LS distribution in the gap immediately from the SCLC CVC. To re-build the narrow or monoenergetical trap distributions (typical for common semiconductors) the most suitable expressions are written with derivatives. The first derivative γ defines the slope of the CVC section in double-log scale relative to the current axis, the η defines the slope of the CVC section in double-log scale relative to the voltage axis. For narrow energy distributions this angle η is too large, and under complete filling-in of the traps it closes to [24-25]. However, it means the slope to the current axis is very small allowing finding the first and higher order derivatives with proper accuracy [44, 45, 48]. It is important that the narrowest trap distributions give the higher accuracy under determination of the derivatives γ, γ', γ'' !

If the distributions in the semiconductor are energetically broadened all expressions (17), (18), and (20) can be used as analytically identical formulas.

As is seen from the expressions written above, in order to receive information about LS distribution three derivatives are to be found at each point of the current-voltage function in various coordinates. Due to experimental peculiarities we had to build the optimization curve as an approximation of the experimental data with it's further differentiation at the sites. The task was solved by constructing smoothing cubic spline [47]. However, the numerical differentiation has low mathematical validity (the error increases under calculation of higher order derivatives). To achieve maximum accuracy we have used the numerical modeling with solving of direct and reverse tasks.

Under solving the direct task we have calculated the functions ρ - E_F and $1/e\ d\rho/dE_F$ - E_F on base of known trap distribution in the gap of the material (the input distribution) using the expressions (12) and (13). Then we have built the theoretical SCLS CVCs ((5), (6)). The mathematical operations are mathematically valid. To solve the reverse problem of the experiment CVCs were worked out using the differential technique based on expressions (17), (21), (18), (20). As a result we have again obtained the deep centers' distribution in the gap of the material (output distribution). Coincidence of the input and output trap distributions was a criterion of the solution validity under solving the reverse task. Further the program set was used for numerical working-out of the experimental CVCs [43-45, 48].

2.2. Determination of deep trap parameters from the functions $1/e \cdot d\rho/dE_F$ - E_F under various energy distributions

Now we determine how the energy position and the trap concentration under presence of the LS in the gap may be found for limit cases by the known dependence $1/e \cdot d\rho/dE_F$ - E_F. In the case of mono-level the LS distribution can be written as $h(E) = N_t\delta(E - E_F)$, where δ is a delta-function.

After substituting this relationship in (12), (13) we obtain

$$n_s \approx n_t = \frac{N_t}{1 + g\exp\left(\dfrac{E_F - E_t}{kT}\right)}, \tag{22}$$

$$\frac{1}{e}\frac{d\rho}{dE_F} \approx \frac{dn_t}{dE_F} = \frac{gN_t\exp(\dfrac{E_F - E_t}{kT})}{kT(1 + g\exp(\dfrac{E_F - E_t}{kT}))^2}, \tag{23}$$

The value of the last function at the maximum ($E_F=E_t$) is $\left(\dfrac{1}{e}\dfrac{d\rho}{dE_F}\right)_{E_{Fm}} = \dfrac{gN_t}{kT(1+g)^2}$,

or at g=1 - $\left(\dfrac{1}{e}\dfrac{d\rho}{dE_F}\right)_{E_{Fm}} = \dfrac{N_t}{4kT}$..

Than

$$N_t = 4kT\left(\dfrac{1}{e}\dfrac{d\rho}{dE_F}\right)_{E_{Fm}}. \tag{24}$$

Thus, building the function $1/e \cdot d\rho/dE_F$ - E_F and finding the maximums by using (24) gives the concentration of discrete monoenergetical levels. The energy position of the maximum immediately produces energy positions of these levels.

If the LS monotonically distributed by energy $h(E)=AN_t=$ const are in the gap of the material it is easy to obtain

$$N_t = h(E) = \dfrac{1}{e}\dfrac{d\rho}{dE_F}. \tag{25}$$

In other words the trap concentration in the sample under such distributions is immediately found from the function $1/e \cdot d\rho/dE_F$ - E_F.

In general case when LS distribution in the gap of the material is described by the arbitrary function their concentration is defined by the area under the curve $1/e \cdot d\rho/dE_F$ - E_F and at low temperatures can be found from the relationship $N_t = \int_{E_1}^{E_2} h(E)dE$. Under reconstruction of such distributions from the SCLC CVC these distributions are energetically broadening depending on the temperature of experiment [43-45]. LS energy positions are again determined by the maximums of the curve.

The correct determination of the trap concentration from the dependence $1/e \cdot d\rho/dE_F$ - E_F may be checked out by using the function ρ - E_F. In case of the mono-level where the Fermi quasi-level coincides with the LS energy position, it is easy to obtain from (22) - $[n_t]_{E_{Fm}} = \dfrac{N_t}{1+g}$,

then $N_t = (1+g)[n_t]_{E_{Fm}}$. If g=1 then $[n_t]_{E_{Fm}} = N_t/2$, $N_t = [2n_t]_{E_{Fm}}$.

If the LS distribution is a Gaussian function ($h(E) = \dfrac{N_t}{\sigma_t(2\pi)^{1/2}}\exp\left(-\dfrac{E-E_t}{\sqrt{2}\sigma_t}\right)^2$) the relationship for determination of N_t is analogous to that described above.

Earlier [43-45] we have described the effect of experimental factors on accuracy of determining parameters of the deep centers by IC method. In Ref. [44, 45, 48] it was shown that the neglecting third order derivative or even the second order derivative does not lead to considerable decrease of the accuracy in determination of the LS parameters. It was demonstrated that under neglecting the 3rd order derivative γ'' in (20) the error in definition of the function $h(E)$ at the point $E_F = E_t$ is no more than 0.4%. At the same time this error is somewhat larger in the interval E_F-E_t~kT but is not larger than (4–7)%. Such a low error of the calculation of the LS parameters is caused by the interception of zero point and the derivative γ'' near the point $E_F = E_t$ (commonly in the range of $|E_F - E_{t_i}| < 0,2kT$). As a result (regarding the absence of accurate experimental measurement of the 3rd derivative) it does not affect the differential working-out of CVCs in the most important section where the Fermi quasi-level coincides with the LS energy position.

If the second order derivative in the working expressions is neglected the error of the defining the function $h(E)$ in the most principal (E_F≈E_t) is about (30-40)%. In both cases the simplification of the expression (21) does not contribute errors to the definition of energy position of the traps' level. Remember that the traditional method of SCLC CVC gives 60-100% error of the traps' concentration [24-25].

3. Methods of preparation and investigation of II-VI films

Thin films CdTe, ZnS, ZnTe were prepared on glass substrates in vacuum by close-spaced vacuum sublimation (CSVS) [49-50]. For further electrical investigations we have deposited hard-melted metal conductive layers on the main substrate by electron beam evaporation (Mo – for CdTe, ZnS; Cr, Ti – for ZnTe). The up-source contact (In(Ag) or Cr in dependence on the conductivity type of the semiconductor) was deposited by the vacuum thermal evaporation. Under condensation of the films of binary compounds the chalcogenide stoichiometric powders were used.

The common temperature of the evaporator was T_e = 973 K for zinc telluride, T_e = (1200÷1450) K for zinc sulfide and T_e = (933÷1023) K for cadmium telluride. The substrate temperature was changed in a wide range T_s = (323÷973) K. Time of deposition was varied: t = (10÷30) min.

Morphology of the samples' surfaces was investigated by optical and electron microscopy. Jeffries' method was used to determine the arbitrary grain size (D) in the condensates. The films' thickness (d) was measured by fractography and interferential methods. The element composition of the layers was studied by X-ray spectroscopy (XRS) analysis using the energy-dispersed X-ray analysis (EDAX) unit or by Rutherford back scattering (RBS) technique (if it was possible). Structural examinations of the films were carried out by the XRD-unit in Ni-filtered K_α radiation of Cu-anode. The XRD patterns were registered in the range of Bragg angles from 20^0 to 80^0. Phase analysis was provided by comparison of inter-plane distances and arbitrary intensities from the samples and the etalon according to the ASTM data [51]. Structural properties of II-VI films are investigated in [20, 49-50, 54-56].

Dark CVC at different temperatures and $\sigma - T$ dependencies of the sandwich-structures (MSM were examined in vacuum by standard techniques (Fig. 1) [21-22].

The power of electronic scheme was estimated by source of stable voltage AIP 120/0.75 that provided a possibility of precise voltage regulation in electric circle in the ranfe of $U = 0.1 \div 120$ V.

A current that passed throught samples in the range of $I = (10^{-9} \div 10^{-5})$ A measured by digital nanoampermeter. Voltage drop on sample was fixed by digital multimeters APPA-108N and UT70B. Sample temperature at inveatigation of electro-physical properties controlled by authomatic feedback temperature controller "OVEN TRM10", input signal from it fed from chromel-alumel thermocouple.

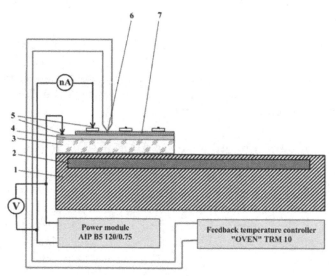

Figure 1. Typical electrical-type scheme for CVC and σ-T characteristic investigations of II-VI semiconductors films: 1 – heater holder; 2 – heater; 3 – glass substrate; 4 – lower conductive layer (Mo, Cr, Ti); 5 – collectors; 6 – thermocouple; 7 – II-VI film

The current mechanisms were identified by the differential method developed in [52-53]. This technique completely analyses j–U, γ–U and $d(\log \gamma)/d(\log U)$-U functions, where γ=$d(\log j)/d(\log U)$ and differentiates satellite and concurrent current mechanisms in the structures and defines the high-field mechanisms among all of them. When the CVCs of multilayered structures were determined by unipolar injection from the source contact the experimental curves were numerically studied by using low-temperature and high-temperature approximations of the IS method [43-45, 48].

PL spectra of CdTe, CdSe and ZnTe films were studied using the spectrometer SDL-1 under excitation of the samples by Ar-laser (λ=514 nm for CdTe and λ=488,8 nm for ZnTe). PL spectra from ZnS films are registered by MPF-4 Hitachi and xenon bulb (λ=325 nm). The temperature in all experiments was stable in the range 4.7÷77 K by using the system "UTREX" [49]. The films CdTe, ZnTe were investigated in the range of edge luminescence, the films ZnS were studied in the impurity energy range.

At interpretation of the PL data it was suggested that the radiation had appeared as a result of electrons' transfer from the conduction (valence) band or shallow donor (acceptor) levels to the deep LS in the gap of the material. Then the activation energy of the processes are defined from the expression:

$$\Delta E = h\nu = E_g - E_i = E_g - (E_a + E_d),\tag{26}$$

where E_a, E_d are energy levels of the donors and acceptors in the gap of the material.

The set of methods for defining parameters of LS in the gap allowed to enhance the accuracy of data obtained and to examine traps and recombination centers with wide energy range.

4. Determination of LS parameters of polycrystalline chalcogenide films by injection spectroscopy method and analysis of $\sigma - T$ functions

4.1. General description of CVC and $\sigma - T$ functions

Dark CVC of sandwich structures current-conductive substrate-film-upper drain contact were measured at different temperatures for examining electrical properties of Zn and Cd chalcogenide films and determination of parameters for LS in the gap of material. Besides that, the function conductivity-temperature was studied in ohmic sections of the CVC and in some cases in the square section of the CVC. Energy positions of donor (acceptor) centers in the films were found from dependencies $\log\sigma = f(10^3/T)$ taking into account their Arrhenius-like character [21-22].

As was shown by the study, the CVC of multilayered structures MSM is defined by the condensation conditions of chalcogenide films, their crystal structure, and material of bottom and upper metallic contacts. CVC of multilayered structures based on low-temperature condensates of II-VI compounds were linear or sublinear. For ZnTe-based MSM structures the CVC were defined by the Pool-Frenkel mechanism, and the data were linearized in the coordinates $\log(I/U) - U^{1/2}$ [52].

Fig. 2 plots typical double-log CVC measured at different temperatures. This figure also shows the function $\sigma - T$ measured at the ohmic section of the CVC.

It is found out that the $\sigma - T$ function of low-temperature condensates are linear with the slope to the T axis decreasing at lowering the measurement temperature. These features are typical for the material with various types of donor (acceptor) impurities with different activation energy. The CVC of high-temperature condensates were somewhat others (Fig. 2). The linear sections are reveled, their slope to the T axis increases as the measurement temperature decreases. It is typical for compensated materials [21-22]. The compensation effect appears more visible under sufficiently low experimental temperatures when the electron concentration becomes close to that of acceptor centers. The slope of the straight lines to the T-axis increases from the value $E_a/2k$ up to the value E_a/k, making it possible to define activation energy for donor and acceptor centers [21-22].

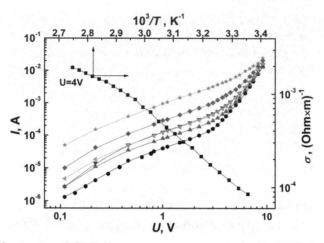

Figure 2. CVC of the structure Cr/ZnTe/Ag at various temperatures: • – T = 298 K; ▲ – T = 303 K; ▼ – T = 308 K; ▶ – T = 313 K; ♦ – T = 318 K; * – T = 323 K, and the dependence $\log\sigma$– $1/T$ obtained from the ohmic section of the CVC. The film is prepared at T_e = 973 K and T_s = 823 K

CVC of multilayered structures where chalcogenide films are prepared at T_s > (500÷600) K were superlinear. As is analytically shown, they are determined by the unipolar injection from the drain contact. Typical SCLC CVCs of the examined films are plotted in Figs. 2-3. CVCs of high-temperature condensates in the range of high field strength a set of linear sections with various slopes to the U-axis was observed. As a rule, the sections with functions: $I - U$, $I - U^2$, $I - U^{3-5}$, $I - U^{8-10}$ were the most pronounced. In some cases after superlinear sections we have observed a square dependence I on U, which had further changed again to the supelinear one with a very large slope η ($\eta \sim$ 13–25). The current jump was revealed and the samples were turn on the low-ohmic state as an irreversible process.

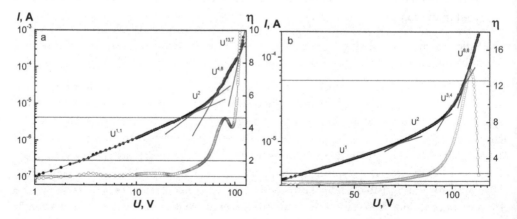

Figure 3. Double-log SCLC CVC of multilayered structures Mo/CdTe/Ag and results of their differentiation. CdTe films are prepared at T_e = 893 K and various T_s: 723 K (a); 823 K (b)

The features of the CVCs are clearly shown in functions $\eta - \log U$ giving a possibility to reveal a fine structure of the CVCs (Fig.3). Each point of this graph defines the slope of the CVC in double-log scale to the voltage axis. Dependencies $\eta - \log U$ were obtained by differentiating the CVC in every experimental point. As it was mentioned above, the problem mathematically reduces to the building smoothing cubic spline which approximates experimental data and its differentiation at the sites.

The curves $\eta - \log U$ resulting from the working-out of the SCLC CVC showed 1-4 maximums in correspondence to the sections of sharp current increase in the $I - U$ dependencies. The most often values of η_{ext} were 8–10. Sometimes the functions $\eta - \log U$ were practically revealed.

Horizontal sections with the almost constant slope $\eta > 2$ were also observed. It may be explained by the presence in the samples of sets of monoenergetical or quasi-monoenergetical levels traps of various energy position and concentration or by availability of the exponential (or other form) LS energy distribution. The specific points of CVCs were used for calculating trap parameters in the material, the ohmic sections helped to find specific conductivity of the layers $\sigma = (10^4 \div 10^5)$ $\Omega \cdot$m. As a result we obtained the concentrational distributions of the traps in the gap of material $h(E)$-E, their energy position (E_t) and concentration (N_t).

At the high voltage the CVCs are typical for the unipolar injection, but, according to [52-53] there some other current mechanisms leading to qualitatively similar current-voltage functions. Thus, we had to identify them additionally according to the procedure described in [52] by analyzing functions $\log I - \log U$, $\eta - \log U$ and $\log \eta - \log U$. It allowed identifying high-voltage current mechanisms in the samples and defining (in some cases) their type.

For further definition of the dominant current mechanism in the base chalcogenide layer we calculated the discrimination coefficient Q_{ext} in the extremum points of the function $\eta - \log U$ and compared it with coefficients typical for other mecahisms [52]. We have found $Q_{ext} > 10^6 \div 10^7$ almost in all cases, what is significantly larger than the values of Q_{ext} typical for the field trap ionization and the barrier –involved current mechanism in the material. This, in turn, points out [52-53] that the extremums in functions $\eta - \log U$ are caused by filling-in of the traps in the material with charge carriers injected from the metallic contact. Using various analytical methods allows to conclude with a good reliability that the CVC's features for multilayered structures with high-temperature chalcogenide layers ($T_s > 500$ K), were caused namely by the SCLC mechanism. Further we have worked out the CVCs due to injection currents only.

Fig. 4. illustrates a typical example of the CVC working-out. It is easy to see that the LS distributions are obtained under analysis of two different CVCs and they are in a good correlation.

To make the distribution more precise we have plotted in the same picture the Gaussian curve. It is seen that for examined polycrystalline CdTe films there are trap distributions in the gap with a form closed to that of the Gaussian one with a small half width σ_t.

Broadening energy levels in CdTe layers prepared by the vacuum condensation may be due to statististical dispersion of polar charge carriers' energy caused by fluctuative irregularities of the film crystalline lattice. This effect is enhanced near the substrate where the most defective layer of the film is grown. This region was an object for determining LS parameters by the method of SCLC CVC.

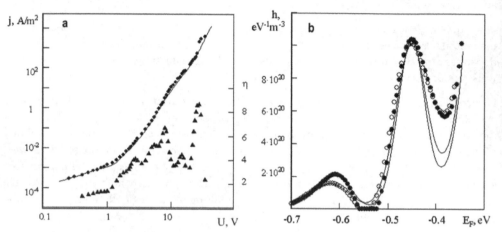

Figure 4. SCLC CVC and its derivative $\eta(U)$ for CdTe-based sandwich structures (a), and trap distribution in the gap of cadmium telluride (b): • – $j(U)$; ▲ – $\eta(U)$ (a); the energy trap distribution is resulted from the high-temperature IS method (b) (• – first measurement; o - repeating measurement at somewhat other temperature); the Gaussian distributions (solid line) are presented for comparison.

4.2. LS parameters from CVC and $\sigma - T$ functions

SCLC CVC was used for determination of trap parameters in the films. The low level of scanning LS spectrum was defined by the position of the equilibrium Fermi level E_{F0}, i.e. its position without charge carrier injection in the sample (ohmic section of the CVC), the upper limit was defined by the position of the Fermi quasi-level at the turn-on of the multilayered structures into low-ohmic state. The start position of the Fermi level was pre-defined by the equilibrium carrier concentration in the material, respectively, by the conductivity of the films. The calculations showed the position of the equilibrium Fermi level E_{F0} was coincide or was close to the energy of the deepest LS in the corresponding samples. The Fermi level is fixed by the traps because the concentration of free carriers in the films is close to the full concentration of LS located at grain boundaries and in bulk crystallites of condensates. As a result, the deepest trap levels located lower than the energy of the equilibrium Fermi level were not revealed in chalcogenide films by the SCLC CVC method.

The possibility of revealing shallow traps in the samples ($E_t \leq 0.21$ eV, for ZnTe films) is restricted by their turn-on into the low-ohmic state stimulated namely by the LS. Thus, the SCLC CVC method had revealed the traps with energy higher positions. However, the traps with different energies also may exist in the samples as shown by the data from the slope of conductivity-temperature functions in ohmic and square sections of the CVCs and luminescence spectra.

4.2.1. CdTe films

Table 1 presents some results of IS calculations for deep centers in polycrystalline and monocrystalline CdTe films. In the gap of the polycrystalline material are LS with E_1 = (0.68÷0.70) eV; E_2 = (0.60÷0.63) eV; E_3 = (0.56÷0.57) eV; E_4 = (0.51÷0.53) eV; E_5 = (0.45÷0.46) eV; E_6 = (0.39÷0.41) eV and concentration N = (10^{18}÷10^{20}) m^{-3}. The concentration of these LS is in the range N_t = (10^{18} ÷10^{21}) m^{-3} and mostly increases with closing their energy positions to the bottom of the conduction band. The traps by the profile $h(E) = N/\sqrt{2\pi}\sigma_t \exp\left(\Delta E^2/2\sigma_t^2\right)$ are similar to the mono-energetical ones with a half width σ_t = (0.011÷0.015) eV. The dominant LS affecting SCLC CVC are the LS with energies E_t = (0.60÷0.63) eV; E_t = (0.56÷0.57) eV; E_t = (0.45÷0.46) eV. Only the traps (if revealed) with E_t = 0.40 eV had the larger concentration.

The LS were registered not only in polycrystalline films but also in monocrystalline layers. We have resolved the traps with E_t = (0.56÷0.57) eV; E_t = (0.52÷0.53) eV; E_t = (0.45÷0.46) eV and E_t = (0.40÷0.41) eV in the gap of the material. The monocrystalline condensates had lower resistance that the polycrystalline layers (10÷100 times), the equilibrium Fermi level in these films was placed more closely to the conduction (valence) band than that in polycrystalline films. Thus, the deepest traps were not revealed by SCLC CVC method in monocrystalline layers. So, the traps $E_t \sim$ 0.70 eV and $E_t \sim$ 0.62 eV found in polycrystalline films may be presented in lower-resistive monocrystalline films.

Ionization energies of the defects in the gap of CdTe were determined from the slope of functions conductivity-temperature in coordinate's $\log\sigma$-1/T [21-22]. Table 2 lists the results for polycrystalline and monocrystalline CdTe films. In high-temperature polycrystalline condensates the following activation energies were observed for conductivity: E_t=0.15; 0.33; 0.40÷0.41; 0.46; 0.60÷0.61, 0.80 eV. In the monocrystalline films the LS had smaller activation energy: E_t=0.06÷0.07; 0.13÷0.14; 0.22÷0.23; 0.29; 0.40; 0.46 eV. Activation energy E_t = (1.50÷1.52) eV is typical for high temperatures of the experiment and corresponds to the gap of the material. The comparison of the LS energy levels from the SCLC CVC and $\sigma - T$ functions is carried out in Table 2. The values E_t from the $\sigma-T$ functions correlate with those observed in CdTe films by SCLC CVC method.

The wide range of the traps revealed in CdTe condensates is obviously caused by investigation of disordered transition layer of the films formed under the film condensation near the substrate. In this layer may be presented foreign impurities adsorbed from the substrate and residual atmosphere under film condensation. Besides that for CdTe the concentration of uncontrolled residual impurities in the charge mixture can be N_t =(10^{20}-10^{21}) m^{-3} which is behind the sensitivity of the IS method. These impurities can form a chain of complexes impurity-native defect producing deep levels in the gap of the semiconductor.

Sample number	d, μm	T_s, K	T_e, K	E_t, eV	N_t, m^{-3}	σ_t, eV
1	8	743	1023	0.63	$4.4 \cdot 10^{19}$	0.030
2 (1st measurement)	19	748	948	0.61	$1.7 \cdot 10^{19}$	0.031
				0.45	$7.3 \cdot 10^{19}$	0.028
2 (2nd measurement)	19	748	948	0.62	$1.5 \cdot 10^{19}$	0.035
				0.45	$8.1 \cdot 10^{19}$	0.032
3	12	748	968	0.68	$7.8 \cdot 10^{18}$	0.023
				0.62	$1.5 \cdot 10^{19}$	0.023
				0.53	$6.1 \cdot 10^{19}$	0.027
4	9	723	893	0.62	$6.6 \cdot 10^{18}$	0.021
				0.56	$4.4 \cdot 10^{19}$	0.016
5	12	823	893	0.62	$2.0 \cdot 10^{18}$	0.023
				0.57	$1.7 \cdot 10^{19}$	0.015
6 (monocrystalline)	11	753	933	0.62	$4.6 \cdot 10^{18}$	0.019
				0.52	$1.3 \cdot 10^{19}$	0.009
				0.41	$1.1 \cdot 10^{20}$	0.016
7	15	753	953	0.60	$2.3 \cdot 10^{18}$	0.019
				0.52	$3.6 \cdot 10^{18}$	0.020
				0.46	$8.6 \cdot 10^{18}$	0.020
				0.41	$1.4 \cdot 10^{19}$	0.015
8	26	758	978	0.61	$3.6 \cdot 10^{18}$	0.023
				0.56	$3.0 \cdot 10^{19}$	0.015
				0.52	$7.4 \cdot 10^{19}$	0.015

Table 1. Parameters of LS revealed in CdTe films by high-temperature IS

E_t, eV			Interpretation
From SCLC CVC	From σ-T dependencies		
Polycrystalline films	Polycrystalline films	Monocrystalline films	
0.68-0.70	0.80	-	V_{Te}^{2+} (0.71 eV [57-61]
0.60-0.63	0.60	-	Te_{Cd}^{2+} (0.59 eV) [57-61]
0.56-0.57	0.57	-	Cd_i^{c2+} (0.56 eV) [57-61]
0.51-0.53	-	-	V_{Te}^{2+} (0.50 eV) [57-61]
0.45-0.46	0.46	0.46	Cd_i^{c+} (0.46 eV) [57-61]
0.39-0.40	0.40÷0.41	0.40	$Te_{Cd}^{2+} \cdot V_{Te}^{+}$ (0.40 eV) [57-61]
-	-	0.29	
-	-	0.22÷0.23	Cd_i^{2+} (0.20 eV) [57-61]
-	0.15	0.13÷0.14	
-	-	0.06÷0.07	

Table 2. Energetical positions of LS levels for defects in the gap of CdTe

As the chalcogenide films were not doped in-advance all LS found here are corresponding to native defects and their complexes with uncontrolled impurities. The interpretation is a challenge while the energy spectrum of PD in the gap of tellurium is studied not enough and identification in the most cases is not satisfactory (Table 3). For example, in [62] the levels E_t of LS are studied by photoinduced currents (PICTS) and authors give more than 150 values of deep levels, where the sufficient part of them is caused by the native defects. More reliables are some theoretical works where energies E_t are calculated («*ab anitio*») [57-61]. We have used namely the data Wei [57-58] obtained from the first principles. Table 3 summarizes our results.

According to calculations the deep centers with energy position 0.71 eV are belonging to V_{Te}^{2+}. We have experimentally observed the level $E_t = (0.68÷0.70)$ eV which may be caused by this defect. Analogically, the LS with energies $(0.60÷0.63)$ eV may be ascribed to the antistructural defect Te_{Cd}^{2+} (0.59 eV), and $(0.56÷0.57)$ eV and $(0.45÷0.46)$ eV to the interstitial cadmium in different charge states: Cd_i^{c2+} (0.56 eV), Cd_i^{c+} (0.46 eV). The level 0.29 eV is also formed by the native defect bound with cadmium Cd_i^{a2+} (0.33 eV). Different ionization energies of interstitial cadmium are due to its place in octo- or tetrahedral position in the crystal lattice of the material.

4.2.2. ZnTe films

Table 4.3 summarizes the results of calculations for ZnTe condensates in dependence on physical technical growth conditions. SCLC CVC method reveals set of trap groups with the most probable energy position $E_{t1} = 0.21$ eV; $E_{t2} = (0.32÷0.34)$ eV, $E_{t3} = 0.57$ eV; $E_{t4} = (0.41÷0.42)$ eV; $E_{t5} = 0.89$ eV. The concentration of the revealed LS is in interval $N_t = (10^{20} ÷10^{21})$ m^{-3}. The LS with energy $E_t = (0.32÷0.33)$ eV are dominant in the most samples and they determine the CVCs of the films.

The trap spectrum in ZnTe films can be partially checked by investigation of temperature – conductivity functions for the condensates. As shown by analysis of $\sigma - T$ functions in the Ohmic section of the CVC for high-temperature ZnTe condensates the following conductivity activation energies are typical: 0.05 eV; $(0.14÷0.15)$ eV; $(0.20÷0.21)$ eV; $(0.33÷0.34)$ eV; $(0.42÷0.43)$ eV; $(0.51÷0.52)$ eV; $(0.57÷0.58)$ eV; $(0.69÷0.70)$ eV and 0.89 eV (Table 4). Set of E_t values from the $\sigma - T$ functions is in a good correlation with those in ZnTe films defined by SCLC CVC method (Table. 3) and low-temperature luminescence (Table. 2).

As the films ZnTe as CdTe layers were not doped in-advance all the calculated LS are due to native PD, their complexes, uncontrolled impurities and their complexes with native defects.

The LS in monocrystals and films ZnTe were studied by SCLC CVC in [24, 63, 64]. Authors [24] have found the trap parameters in monocrystalline samples by the voltage of complete trap filling-in: $E_t = 0.17$ eV and $N_t = 10^{22}$ m^{-3}. On the other hand, measurements of $\sigma - T$ dependencies in the square section of the CVC gave $E_t = 0.14$ eV and $N_t = 10^{23}$ m^{-3} taking in mind the presence of traps in the material authors [64] calculated the LS density in ZnTe films prepared by laser evaporation technique: $N_t = (4.2 \div 8.4) \cdot 10^{22}$ m^{-3}. The trap energy is not defined in this work. In ZnTe films obtained by the electro-deposition the trap concentration is $N_t = 3.6 \cdot 10^{21}$ m^{-3} [63].

Parameters	From SCLC CVC		From σ-T dependencies	
of film condensation	E_t, eV	N_t, m^{-3}	E_t, eV	N_t, m^{-3}
$T_s = 623$ K,	-	-	0.21	$2.1 \cdot 10^{20}$
$T_e = 973$ K	0.34	$8.6 \cdot 10^{20}$	0.34	$7.3 \cdot 10^{20}$
$T_s = 673$ K, $T_e = 973$ K	0.33	$5.3 \cdot 10^{20}$	-	-
$T_s = 723$ K, $T_e = 973$ K	0.34	$2.9 \cdot 10^{20}$	0.33	$4.1 \cdot 10^{20}$
	-	-	0.57	$5.5 \cdot 10^{20}$
	-	-	0.89	$8.4 \cdot 10^{20}$
$T_s = 773$ K, $T_e = 973$ K	0.32	$5.3 \cdot 10^{20}$	-	-
$T_s = 823$ K, $T_e = 973$ K	0.42	$2.1 \cdot 10^{20}$	-	-
$T_s = 873$ K $T_e = 973$ K	0.32	$1.5 \cdot 10^{21}$	-	-
$T_s = 623$ K $T_e = 973$ K	0.35 0.37	$8.8 \cdot 10^{20}$	-	-

Table 3. LS parameters in ZnTe films from SCLC CVC and σ-T functions

As is seen from the Table 3, the trap concentration in ZnTe films is significantly lesser than that in condensates prepared by laser evaporation, electro-deposition methods and even in the monocrystalline material [23, 63- 64]. It shows a high structural perfectness and stoichiometry of the layers.

Nevertheless, the most levels found in ZnTe films may be identified with some probability. The level $E_1 = 0.05$ eV is commonly bound with single-charged dislocation V_{Zn}^-, and the level $E_2 = 0.15$ eV is bound with a double-charged V_{Zn}^{2-} Zn vacancy [65, 66]. In later works the second level is ascribed to Cu as to a traditional residual impurity in ZnTe, and the double-charged Zn vacancy is supposed to have a more deeper energy level 0.21 eV [66]. It is thought that the energy activation $(0.36 \div 0.40)$ eV [67, 68] is for the common substitution impurity in ZnTe, namely O_{Te}. The most deepest level 0.58 eV authors [67] ascribe to the Te vacancy V_{Te}^{2+} (interstitial zinc Zn_i^{2+}). The possible interpretation of LS in ZnTe films is listed in table 5.7. Other energy levels on our opinion are belonging to the uncontrolled impurities and complexes native defect-impurity.

We have revealed trap levels with energy position $E_t = (0.22 - 0.25)$ eV and concentration $N_t = (5.0 \cdot 10^{20} \div 1.5 \cdot 10^{21})$ m^{-3} by the analysis of SCLC CVC in ZnS films. These LS may be localized in the gap due to presence of the interstitial Zn atom Zn_i^{2+}. LS with energy position $E_t = (0.24 \div 0.25)$ eV were observed in [70] the thermo stimulated current technique.

T_s, K	623		673	723		773	823	873	Interpretation
Section E_t, eV	Ohm.	Sq.	Ohm.	Ohm.	Sq.	Ohm.	Ohm	Ohm.	
E_9						0.89			–
E_8	0.70							0.69	–
E_7			0.58	0.58	0.57	0.58			$V_{Te}^{2+}(Zn_i^{2+})$ [67]
E_6	0.51			0.52		0.51	0.52		V_{Zn}^{2-} (0.50) [69]
E_5	0.43		0.42				0.42		O_{Te} (0.41) [67]
E_4	0.33	0.34	0.33			0.33	0.34	0.33	O_{Te} (0.36) [68]
E_3		0.21		0.20		0.20			V_{Zn}^{2-} (0.21) [67]
E_2						0.14		0.15	V_{Zn}^{2-}, Cu_{Zn}^{+} (0.15) [69]
E_1		0.05							V_{Zn}^{-} (0.05) [65]

Table 4. Energy positions of LS for defects in ZnTe gap determined from the slope of σ–$1/T$ functions

According to the Arrhenius equation σ-T functions allowed calculating conductivity activation energies in linear sections: E_1=0.03 eV; E_2 = (0.07÷0.08) eV, E_3 = 0.15 eV; E_4 = (0.23÷0.24) eV; E_5 = 0.33 eV; E_6 = 0.46 eV; E_7 = 0.87 eV.

4.2.3. ZnS films

Table 5 summarizes LS parameters calculated by SCLC CVC method and from the σ-T functions in ZnS condensates prepared under various physical technical conditions. Reference data are presented for comparison. The table shows a correlation between our results and data obtained by other authors [71-74]. Besides that, there is a coincidence of defect energy positions defined from the SCLC CVC and σ-T functions.

T_s, K (T_e = 1173 K)	From CVC SCLC		From σ-T dependencies	
	E_t, eV	N_t, m^{-3}	E_t, eV	
				Reference data
523 K	0.22	$5{,}1 \cdot 10^{20}$	0.03	0.029 [71]
			0.07	0.06 [72]
590 K	0.25	$1.5 \cdot 10^{21}$	0.07	0.06 [72]
			0.24	0.24; 0.25 [70, 72]
			0.33	0.31-0.33 [70]
			0.87	0.81-1.29 [74]
673 K	0.23	$8{,}2 \cdot 10^{20}$	0.03	0.029 [71]
			0.23	0.24; 0.25 [70, 72]
			0.15	0.14 [72, 73]
			0.46	–

Table 5. LS parameters defined by analysis of SCLC CVC and σ-T functions at ohmic section of the CVC for ZnS films prepared under various physical technical condensation modes

All the LS found here were not identified because of absence of corresponding reference data. Only the levels with activation energy $E_1 = 0.15$ eV and $E_2 = (0.22 \div 0.25)$ eV may be bound with single- Zn_i^+ and double charged Zn_i^{2+} interstitial Zn atom.

5. Determination of LS parameters of polycrystalline chalcogenide films by optical spectroscopy (low-temperature photoluminescence)

Low-temperature photoluminescence (PL) is one of the most reliable tools applied for investigation of longitudinal, native, impurity and point defect ensembles in semiconductors. High resolution of the method makes it possible to examine not only bulk materials (bulk chalcogenide semiconductor are now good studied [75-87, 90-93, 96, 99-100, 102-104]), but also thin films, in particular, chalcogenide semiconductor thin layers. In this part we present data obtained by studying low-temperature PL spectra of ZnTe, CdTe and ZnS films. These results allowed monitoring and adding new results to those given by the IS method.

5.1. CdTe films

Fig. 5 (a, b) illustrates the typical spectra of these films. As shown, the spectra for both types of the films are significantly similar. A modest energetical displacement of lines in spectra from epitaxial films comparing to those from the polycrystalline layers films deposited on glass may be caused by presence of sufficient macrodeformations in the layers CdTe/BaF$_2$. PL spectra from CdTe layers have lines originated from optical transfers with participation of free and bound excitons, transfers valence band – acceptor (e-A), donor-acceptor transfers (DAP), the radiation caused by presence of dislocations or DP (donor pairs, DP) (Y - stripes); the spectra also have a set of lines corresponding to optical transitions where phonons take place (LO - phonon replica) [87-99].

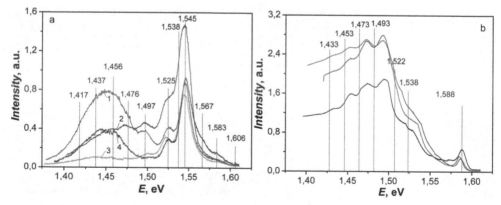

Figure 5. PL spectra registered at $T=4.5$ K for polycrystalline films CdTe/glass (a), prepared at $T_e = 893$ K and various T_s, K: 473 K (1); 523 K (2); 623 K (3); 823 K (4) and for the epiraxial layer CdTe/BaF$_2$ (b): $T_e = 893$ K, $T_s = 798$ K

Activation energies relative to the valence band (while the most samples were of p-type conductivity) were calculated using expression (26) (in analogy with description above).

The gap of CdTe at T=4.5 K was supposed to be E_g =1.606 eV. The data are presented in Table 6.

E_i, eV	Reference data, eV	ΔE_i, eV	Recombination type	Interpretation
1.583	1.589-1.588	0.023	Exciton	A^0X, A -Li, Na [90, 92, 95, 99, 100]
1.567	1.568	0.039	Exciton	(A^0X)-LO [90, 100]
1.545	1.546	0.061 (0.050) [13] (0.058-Li. Na)	e-A - - DAP	A-V_{Cd}^- [13, 94, 95] $A - Li$, Na [90, 92, 99] $(V_{Cd} - O_{Te})^-$[97, 102] $D - V_{Cd}^-$ [13, 103]
1.538	1.538	0.068 (0.067-V_{Cd})	DAP -	$D - A$ [93] $D - A$ (Na) [13, 92]
1.525		0.081	(e-A) LO	[97]
1.497	1.496 1.499 1.495	0.109 (0.107 [13]) (0.111 [13])	e-A - -	O [99] Ag_{Cd} [13] V_{Cd}^{2-} [13]
1.476	1.474 1.477	0.130	e-A	Y [96-99, 101] Y (α-dislocations [93]
1.457	1.459	0.149 (0.146 [13, 92])	e-A - - DAP	$(A$-$X)$-LO [101] Y [91] Cu_{Cd} [13] $\left(In - V_{Cd}^{2-}\right)^-$ [13, 92, 100]
1.438	1.436	0.168	e-A DAP	$(A$-$X)$-2LO [101] $\left(In - V_{Cd}^{2-}\right)^- - LO$ [13, 92, 100]
1.419	1.415	0.187	e-A DAP	$(A$-$X)$-3LO [101] $\left(In - V_{Cd}^{2-}\right)^- - 2LO$ [13, 92, 100]

Table 6. Principal lines of PL spectra of CdTe films and their interpretation

The lines due to exciton recombination in CdTe single crystals are well-known. Authors [13] show energy level diagram for the exciton localized on neutral donors or acceptors and possible transfers between these levels. Commonly the elements of 3rd Group (Ga, In, Al) and 7th Group (Cl, Br, I) are shallow donors in CdTe and ZnTe, and acceptors are the elements of 1st Group and 5th Group (Li, Na, Cu, Ag, Au, N, P, As). These elements are typical excessive impurities in compounds II-VI. Authors [13] also give the ionization energies of principal dopant impurities in CdTe: for the donors (13.67÷14.48) meV, for acceptors 56 (N) - 263 (Au) meV. We have used these values for further interpretation of the experimental results.

Unlike ZnTe condensates the peak bound with a free exciton recombination at energies E_i = 1.596 eV [12] had not been observed for CdTe films. However, the spectra showed a line caused by the recombination of exciton localized on neutral acceptor A^0X - E_i = (1.583÷1.588) eV (1.589 eV [90, 92, 95, 99, 100]). This line indirectly demonstrated that the investigated films were of p-type conductivity and correspondingly low concentration of dopant impurities. Maybe there is a reason for absence of the peak bound with the exciton localized on the neutral donor D^0X – 1.593 eV [13, 90, 100] in the registered spectra. The excessive impurity (Li, Na) commonly is an acceptor in II-VI compounds which produces shallow LS near the valence band.

In some PL spectra of CdTe films we have observed the peak due to the phonon repetition of the line from the bound exciton (A^0X)-LO at E_i = 1.567 eV. The similar peak with E_i=1.568 eV and E_i=1.570 eV was also observed in [90, 100]. It should be noted that the excitation energy of the longitudinal phonon in CdTe is LO(Γ) – 21.2 meV [13, 90, 92]. This value is almost coinciding with that observed experimentally (21 meV) showing our correct interpretation of the experimental data.

The most intensive peak of 1.545 eV was observed in PL spectra from polycrystalline films. The similar peak with energies E_i=1.55 eV and E_i=1.545 eV was registered by authors [92, 94, 97, 99, 103]. The common interpretation says that this peak is caused by the electron transition between the conduction band and acceptor (e-A) (a single-charged vacancy V_{Cd}^- [94] or other shallow acceptor [92, 99]). Nevertheless, authors [13, 95, 103] point out this radiation as a consequence of p resenting donor-acceptor pairs (DAP) where the acceptor is a native defect (V_{Cd}^-) [13, 103] or another uncontrolled shallow impurity [95]. Authors [97] have found the activation energy of corresponding donors and acceptors: 8 meV and 47meV.

Results of investigated polycrystalline CdTe films in hetero structures CdTe/ZnS under air and vacuum annealing have given [97] another interpretation. It is supposed that the luminescence with 1.55 eV is due to oxygen presence in the material. In which form it exists in the material (substitutional impurity or oxide phase) is not established. However, authors [102] have studied the LS in CdTe single crystals by thermo electronic spectroscopy and demonstrated the energy level 0.06 eV bound with a complex ($V_{Cd} - O_{Te}$).

Analyzing our results allows us to conclude that the peak E_i=1.545 eV is rather due to the electron transitions between the conduction band and acceptor (a single-charged vacancy or DAP). Really if this peak was caused by oxygen we could it observed in PL spectra from both polycrystalline and epitaxial films but there are no such a peak in PL spectra from the films CdTe/BaF$_2$. Besides that no structural method had revealed the oxygen in these compounds The films under investigation have shown no registered donor impurities of considerable concentration, so the interpretation of this peak as a consequence of the DAP presence is lesser probable than that a consequence of e-A transition.

In some cases the PL spectra from the polycrystalline films showed an asymmetric peak 1.545 eV indicating that in reality it may be a superposition of two nearest lines. Mathematical analysis showed that the most probable position of the additional peak is E_i = 1.538 eV. The similar peak was observed in spectra from the epitaxial films CdTe/BaF$_2$. The line with the same energy was revealed by authors [93] in PL spectra from deformed CdTe single crystals and is supposed to be caused by defects generated in the material due to slip of principal Cd(g)-dislocations. Authors [13, 92] explain the peak E_i= 1.538 eV as one of unknown nature. Similar interpretation is also in [89] where the line E_i= 1.539 eV is caused by DAP (here the acceptor is sodium, Na_{Cd}). The next peak E_i = 1.525 eV is likely is the phonon repetition of the previous one (e-A)-LO [96].

The PL line E_i = 1.497 eV was observed in [99] on monocrystalline CdTe samples under doping by ion implantation. As this line has appeared in the samples doped with oxygen only the authors suggest it is caused by the presence of this impurity. Other authors suppose this line is due to electron transitions between the conduction band and the level of the substitutional impurity acceptor Ag_{Cd} (E_v+0.107 eV) [13] or by native defect V_{Cd}^{-2} (E_v+0.111 eV) [13].

The wide radiation stripe in polycrystalline films at the energy 1,45 eV is separated in single peaks based on results from the PL of epitaxial films. They are shown in Fig. 5.

The peak 1.476 eV in [96-98] is due to longitudinal defects (dislocations and DP, a so-called Y-stripe). Authors [90, 98] assume the Y-stripe at (1.46-1.48) eV is caused by longitudinal defects (dislocations). Authors [99] make it more precicely: this peak is caused by the recombination of exciton localized on slipped Cd-dislocations. Authors [93] have investigated The photoluminescence of deformed CdTe single crystals and showed that the peak E_i=1.476 eV is not caused by the Cd-dislocations but is due to the electron states of 60^0 Te(g)-dislocations (α-dislocations). So that, the number of authors have the same opinion that this line in PL spectra is caused by the longitudinal defects. We also agree with this interpretation.

The lines 1.453 eV, 1.433 eV and 1.413 eV which are good resolved in spectra from the epitaxial films CdTe/BaF$_2$ are very similar to 1LO, 2LO, 3LO repetitions of the peak E_i = (1.473÷1.476) eV. However, the energy difference of these lines (ΔE=0.0200 eV) does not coincide with the energy of longitudinal optical phonons in CdTe 0.0212 eV making it difficult to interpret the corresponding peaks unambiguously. At the same time, the analogous set of lines with the LO structure and the energy difference 0.0200 eV in the range ΔE = (1.39÷1.45) eV was observed by authors [101]. They have studied polycrystalline CdTe films deposited by vacuum evaporation at T_s= (723÷823) K on glass and aluminum substrates.

Authors [100] have examined undoped and doped with donor impurities (Al, In) CdTe single crystals and also have observed the PL stripe in the energy range ΔE = (1.380÷1.455) eV containing four lines with LO structure. The authors interpreted them as electron transition between DAP and their phonon repetitions. Authors [81, 91] suppose the wide peak 1.46 eV is due to the excitons localized in longitudinal defects, probably dislocations

(*Y*-stripe). The lines 1.455 eV, 1.435 eV and 1.415 eV were observed in [94] from the polycrystalline CdTe films prepared by the gas-transport method.

As we see the most authors have an unique opinion: the set of lines in the range ΔE=(1.413÷1.476) eV is due to longitudinal defects (rather dislocations), and their intensity [93] can be a measurement unit of these defects in the material.

For polycrystalline films (Fig. 5, a) the LO structure of the stripe caused by the longitudinal defects at energies ~1.45 eV has practically not been observed, maybe because of superposition with additional lines of another origin.

The defect complexes in the material (A-centers) are also resolved by the PL in the same energy range (it can be considered as a partial case of DAP). According to [13, 97] A- centers $\left(V_{Cd}^{2-} - D^{+}\right)^{-}$ where Cl is a donor produce the line and its LO-phonon repetitions with energies 1.454, 1.433, 1.412, 1.391, 1.370, 1.349 and 1.328 eV. However, as is seen in Fig. 5, this stripe is displaced relatively to that observed experimentally, so the experimental PL spectra of CdTe films can be completely explained by these complexes only. The narrower stripe with peaks 1.458, 1.437, 1.417 and 1.401 eV produces the A-center where indium is a donor. This stripe has the better coincidence with experimental one but is also displaced. Besides that, it is difficult to explain why the A-complex is observed in the polycrystalline films and is not observed in the epitaxial layers while the charge mixture for both types of the films is the same. Thus we suppose the interpretation of the wide stripe in the energy range ΔE = (1.413÷1.4760) eV due to longitudinal defects is more reliable.

Under change of condensation conditions of polycrystalline samples we have observed the change of intensity for a stripe due to prolonged defects (~1.45 eV). As shows Fig. 5, as the substrate temperature increases from 473 K to 623 K the intensity of this stripe is decreasing and then it increases as the T_s increases. These results have a good correlation with data of investigation of CdTe film substructure [49], this fact points out an enhance of the structural quality (lowering vacancy concentration) of the bulk crystallites in condensates under elevating substrate temperature up to T_s=623 K, but this quality becomes lower as the substrate temperature increases over 623 K.

As the substrate temperature elevated (T_s>723 K) the optical properties of CdTe films were strongly degraded forming a number of additional peaks in the PL spectra which finally become a bell-like curve without possibility to identify the separate lines. Morphological studies demonstrated further increase of the crystallite sizes in this temperature range. However, the volume of these crystallites becomes a high-defective one.

Table 6 summarizes results of PL spectra interpretation for CdTe films showing their high optical quality.

5.2. ZnTe films

Fig. 6 illustrates typical PL spectra of ZnTe films registered at 4.5 K. A number of lines is observed, their energies are indicated in the Fig. 6 and are listed in Table 7. Analysis and interpretation of the PL peaks are carried out according to reference data [75-89].

The low-temperature PL spectra of ZnTe films show a set of peaks originated from: i) optical transitions under participation of free (X) and bound on neutral donor (D^0X) and acceptor (A^0X) excitons; ii) transitions valence band – acceptor impurity (e-A), iii) radiation due to presence of longitudinal defects (dislocations, Y-stripe); iv) optical transitions where phonons of different type are participating (LO (0.0253 eV), TO, LA (0.0145 eV), TA (0.007 eV) -repetition).

We calculated activation energies of corresponding processes using the expression (26). The gap of ZnTe crystal at 4.5 K was supposed to be E_g = 2.394 eV. As the examined material was of p-type conductivity the activation energies were counted down relative to the valence band. Table 7 summarizes these data.

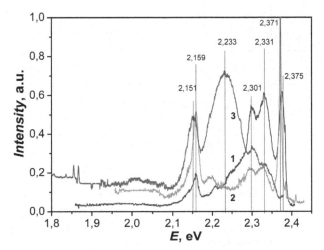

Figure 6. Photoluminescence spectra registered at T=4.5 K for ZnTe films prepared at T_e = 973 K and various T_s, K: 573 K (1); 673 K (2); 773 K (3)

Optical transitions with energy (2.381÷2.383) eV were observed in [68, 75-82, 84-86] where authors had studied monocrystalline or bulk polycrystalline ZnTe of high structural and optical quality. These transitions are commonly relating to a free exciton (X). Earlier [82] the PL line of E_i = (2.374÷2.375) eV was suggested to be caused by the exciton bound on neutral acceptor (zinc vacancy V_{Zn}). Further [76-78] it was shown that other acceptor centers take part in forming such an excitonic complex, in particular, acceptor centers due to uncontrollable impurities (Li, Cu) in ZnTe are of interest. However, in the most recent works [68, 81] this line is ascribed to the exciton localized on shallow neutral donor (atoms of uncontrollable impurities from 3rd and 7th Groups of the Periodical System (In, Ga, Al, Cl, Br. I)). These impurities form in the gap of the material more narrower levels than the acceptor ones. The line with E_i=2.371 eV which is energetically closed to that considered above is due to radiation of bound excitons [76-78, 81]; nevertheless the impurity (acceptor) in this complex has (obviously) somewhat larger energy level causing other energy of the stripe. These acceptors are native defects and uncontrollable excessive impurities (Li, Na, Ag, Cu).

According to [81] Li_{Zn} is the most probable candidate while it forms in the gap of the material energy level 60.6 meV.

Radiation line E_i, eV	Reference data, E_i, eV	Activation energy, ΔE, eV	Recombination type	Interpretation
2.383	2.381÷2.383	0.011	Exciton	X, n=1 [68, 76, 82]
2.375	2.375 2.379	0.019	Exciton	A^0X [76-77] A- V_{Zn} [82] D^0X, D- In [68, 86]
2.371	2.374; 2.375	0.023	Exciton	A^0X, A-Li,Cu [68]
2.331	2.334; 2.332	0.060 (0.061 – Li [75, 78]) (0.063 – Na [75])	e-A	A - Li_{Zn}, Na_{Zn} [75, 76]
2.301	2.307	0.093	(e-A)-LO	A - Li_{Zn} [76-77]
2.270	2.270	0.124	e-A	A - Ag_{Zn} [76]
2.233	2.230	0.161	e-A	A - Cu_{Zn} [75, 79]
2.208		0.186		
2.194	2.195; 2.19	0.200	e-A	Y_1 [75]
2.159		0.235		
2.151	2.155	0.243	e-A	Y_2 [75, 76, 78, 84]

Table 7. Principal lines in PL spectra of ZnTe films and their interpretation

It should be noted that the presence of excitonic lines in PL spectra from high-temperature ZnTe condensates points out their high optical and crystal quality. These lines are of sufficient intensity in the spectra from the films deposited at the substrate temperature T_s=573 K and the larger intensity for condensates prepared at T_s=673 K. Excitonic lines in the spectra from low-temperature condensates and layers manufactured at T_s>773 K were not registered. Thus, the results of PL studies indicate that the films deposited at the substrate temperatures T_s= (623÷673) K are the most optically perfect layers. These data are coinciding with the results of investigations of substructural characteristics of ZnTe films reported earlier [54]. According to these data the dependence of the CSD sizes on the substrate temperature is a curve with the maximum at T_s = (600÷650) K. The minimal dislocation concentration is also observed in the films under these temperatures.

The line E_i = 2.34 eV belonging [66] to V_{Zn} is not observed in spectra of the radiation recombination in ZnTe films. This fact is also confirming high stoichiometry of the films under study.

The set of nearest lines in the energy range ΔE = (2.30÷2.33) eV and ΔE = (2.17÷2.25) eV authors [76-79] ascribe to the electron transitions from the conductance band to the shallow acceptor levels formed by Li or Cu atoms and their phonon repetitions (LO – 25.5 meV). There are stripes $2SLi$, $3S_bLi$, (e-A) Li, $2P$ Li, $4S_bLi$, $4S_bLi$-LO, $2SCu$, $3S_bCu$, $4S_bCu$, $2S_b$ Cu - LO, $2S_b$ Cu - 2LO and others. Experimental and theoretical values of the activation energy for ground and excited states for the main excessive impurities in ZnTe (lithium and copper) are reported in [76]. They are in the energy range ΔE = (0.0009÷0.0606) eV for Li and ΔE = (0.001÷0.148) eV for Cu. However, in [65] the line E_i = 2.332 eV is supposed to be due to other excessive impurity Na_{Zn}, and in [82] this line is due to the native defect V_{Zn}. Another optical transition E_i = 2.27 eV authors [77] ascribe to the Ag impurity 2S Ag.

What about the peaks in the energy range E = (2.10÷2.21) eV. These transitions were for the first time observed in [75-79] and authors had called them Y_i-lines. They are ascribed to the distortions of the crystalline lattice of the material near incoherent twin boundaries, dislocations and other longitudinal defects where the dangling bonds are formed in the semiconductor material. So that, the lines E_i = 2.159 eV and E_i=2.194 eV can be interpreted as Y_2 (2.155 eV) and Y_1 (2.195 eV) [75]. They are due to longitudinal defects and the change of their intensity may point out the change of these defects concentration in the material. Somewhat other energy position of the line due to oxygen (2.06 eV) is reported in [66]. Thus, analysis of the reference data has forced us to conclude that PL lines in the energy interval ΔE = (1.835÷2.055) eV are rather caused by oxygen, its complexes and phonon repetitions. If it is true, the analysis of PL spectra from ZnTe films indicates the increase of the oxygen content in the samples under increasing the condensation temperature. Actually, if there is no oxygen in the samples prepared at 573 K, its concentration in high-temperature films (T_s=773 K) is sufficiently larger. Oxygen concentration in the material strongly depends on the vacuum conditions under the film preparation and the charge mixture quality.

5.3. ZnS films

Low-temperature photoluminescence is the most reliable tool for examining wide gap materials providing minimization of overlapping peaks due to various recombination processes. The typical PL spectra from ZnS films at 4.7 K are shown in Fig. 7. The detailed analysis of the PL spectra (identification of complex broadened lines) was carried out by ORIGIN program. Maximums of the peaks revealed by this analysis (Fig.7) are noted by vertical lines.

It should be noted that the PL spectra registered at various temperatures of experiment have no sufficient distinctions except those with somewhat larger line intensities in spectra obtained at 77 K. Analysis of the spectra shows that for ZnS films deposited at T_s = (393-613) K the peaks with λ_i = 396 nm (E_i = 3.13 eV) and λ_i = 478 nm (E_i=2.59 eV) are dominating. Further working-out of the spectra demonstrated that the peak λ_i= 396 nm is asymmetric (Fig. 7) what is may be explained by the superposition of two closely placed lines. The spectra also have low intensity peaks with λ_i=603 nm (E_i=2.06 eV) and λ_i=640 nm (E_i=1.94 eV).

PL spectra from the films prepared at higher T_s is sufficiently changed. There is a number of overlapping peaks where the most intensive ones are in the wavelength range $\Delta\lambda_i$ = (560÷620) nm.

Under interpretation of PL spectra from ZnS films we have calculated the activation energies of processes causing the corresponding lines. We also have suggested the PL radiation took place under transfers of electrons from the conduction band (or shallow donors) to the deep LS in the gap of the material. Then the optic depth of the energy level of the defect (ΔE) relative to the valence band causing the spectral peak may be found from (26) supposing the optical gap of the material at 4.5 K is $E_g = 3.68$ eV.

Taking into account that the chalcogenide films were not doped in-advance one can suggest that the lines in spectra are due to transfers of carriers between conduction band and LS caused by the native point defects, their complexes and uncontrolled impurities. We made an attempt to identify these LS according to reference data [104-108] (Table 8). As is shown there is a good correlation of our results and those obtained by other authors for ZnS single crystals.

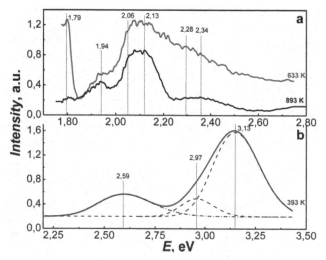

Figure 7. Typical PL spectra for ZnS films (a) and the example of the peak differentiation (b)

The investigations have shown that the Schottky defect V_{Zn} is a dominant defect type in ZnS films prepared at low substrate temperatures $T_s = (393\text{-}613)$ K. As T_s increases the number of single-charged Zn vacancies in the condensates decreases, and concentration of double-charged Zn vacancies increases. In the films deposited at higher substrate temperatures $T_s = (653\text{-}893)$ K single-charged S vacancies V_s^+ and double-charged S vacancies V_s^{2+} and interstitial Zn atoms Zn_i are dominating.

Such features of the PD ensemble in the samples are obviously caused by processes of condensation and re-evaporation of Zn and S atoms from the substrate. Actually, at low T_s the defect formation in the films is determined by higher S pressure comparing to Zn pressure in the mixture vapor providing Zn vacancy formation in ZnS condensates. As T_s increase the PD ensemble in the material is determined by the more rapid re-evaporation of the same S atoms from the substrate resulting in production of Zn-beneficiated films. Sulfur vacancies and interstitial Zn atoms are being dominant defects in such condensates.

T_s, K	Measurement range λ, nm	λ_1, nm E_1, eV	λ_2, nm E_2, eV	λ_3, nm E_3, eV	λ_4, nm E_4, eV	λ_5, nm E_5, eV	λ_6, nm E_6, eV	λ_7, nm E_7, eV	λ_8, nm E_8, eV
393	360÷640	**396**	417	478	-	-	603	-	-
		3.13	2.97	2.59	-	-	2.06	-	-
613	450÷720	-	-	478	-	-	603	640	-
		-	-	**2.59**	-	-	2.06	1.94	-
653	450÷720	-	-	-	530	**582**	603	640	**690**
		-	-	-	2.34	**2.13**	**2.06**	**1.94**	**1.79**
893	450÷720	-	-	-	530	**582**	603	640	690
		-	-	-	2.34	**2.13**	**2.06**	1.94	1.79
ΔE_i, eV Exper.	-	**0.55**	**0.71**	**1.09**	**1.34**	1.55 **1.40**	1.62 **1.40**	**1.74**	1.89 **1.74**
Defect	-	V_{Zn}^-	$D_s, V_{Zn}^{2-})S_i^-$	V_{Zn}^{2-}	Cu	V_S^{2+}	V_S^{2+}	V_S^+	V_S^+
ΔE_i, eV ref.	-	0.60	0.70	1.10	-	1.40	1.40	1.90	1.90
Reference data	-	[107]	[107, 108]	[104]	[104]	[104]	[104]	[104]	[104]

Table 8. Results of PL spectra working-out and their comparison with reference data (solid values are for peaks of maximum intensity)

The PL spectra of ZnS films have also revealed low intensity lines from the activator impurity (Cu) and, possible, S_i^- [106] or a complex defect (O_s, V_{Zn}^{2-}) [107]. Results of studying low-temperature PL allowed constructing energy position model of native point defects in zinc sulfide films prepared by quasi-closed space technique (Fig. 8).

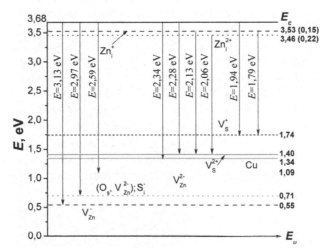

Figure 8. The model of the level positions for native point defects in band gap of ZnS films

To explain the experimental results we have used quasi-chemical formalism for modeling the point defects ensemble in the examined chalcogenide films in dependence on physical technical conditions of layer condensation. This method concerns all defects, electrons and holes as components of thermodynamic equilibrium in the bulk crystal (complete equilibrium of the point defects). Then the modeling procedure reduces to solving set of equations which describe penetration of point defects into solid from the gas state along with the equation of electroneutrality and intrinsic conductivity equation [109-110]. The most complete spectrum of the native defects was taken into account under modeling the point defects ensemble. Calculations were carried out for the complete defects' equilibrium as well as for their quenching. Under modeling we have used energies of native defects formation obtained «*ab initio*» in [56-60]. Reference data of ionization energies of acceptor and donor centers of point defects in CdTe, ZnS, and ZnTe were used along with results of our experiments. Main data of modeling are presented in [111-116].

6. Conclusions

1. Express-method of IS providing maximum information on deep centers in high-resistive films based on analysis of SCLC CVC is developed and allows without additional studies
- identifying current mechanism in the structure as SCLC;
- receiving correct information on LS parameters in the gap of material: energy position, concentration and energy distribution immediately from the experimental CVCs without model framework.
2. Using IS method the LS spectrum in the gap of polycrystalline (and monocrystalline) films of II-VI compounds is examined. These results are checked and made more accurately by analysis of σ - T – functions and low-temperature luminescence.
3. Using the methods mentioned above in the gap of polycrystalline material are revealed the LS with following energy positions: $E_{t1} = 0.05$, $E_{t2} = (0.14 \div 0.15)$, $E_{t3} = (0.20 \div 0.21)$, $E_{t4} = (0.32 \div 0.34)$, $E_{t5} = (0.42 \div 0.43)$, $E_{t6} = (0.51 \div 0.52)$, $E_{t7} = (0.57 \div 0.58)$, $E_{t8} = (0.69 \div 0.70)$ eV (ZnTe); $E_{t1} = (0.13 \div 0.15)$, $E_{t2} = (0.39 \div 0.40)$, $E_{t3} = (0.45 \div 0.46)$, $E_{t4} = (0.51 \div 0.53)$, $E_{t5} = (0.56 \div 0.57)$, $E_{t6} = (0.60 \div 0.63)$, $E_{t7} = (0.68 \div 0.70)$ eV (CdTe); $E_{t1} = 0.03$, $E_{t2} = (0.07 \div 0.08)$, $E_{t3} = 0.15$, $E_{t4} = (0.23 \div 0.24)$, $E_{t5} = 0.33$, $E_{t6} = 0.46$, $E_{t7} = 0.87$, $E_{t8} = 1.94$, $E_{t9} = 2.34$, $E_{t10} = 2.59$, $E_{t11} = 2.97$, $E_{t12} = 3.13$ eV (ZnS) and concentration $N_t = (10^{19} - 10^{21})$ m^{-3}. Comparing reference data produced an identification of these levels as ones belonging to native point defects, uncontrolled impurities and their complexes. The wide range of LS revealed is due to high-sensitive methods used under investigations as well as because of examining traps in the intermediate layer of the films forming under condensation near the substrate.

Author details

Denys Kurbatov and Anatoliy Opanasyuk
Sumy State University, Sumy, Ukraine

Halyna Khlyap
TU Kaiserslautern, Kaiserslautern, Germany

Acknowledgement

This work is supported by the Ukraine State Agency for the Science, Innovation and Informatization and by the NRF grant funded by the MEST of Korea within the project «Advanced materials for low-cost high-efficiency polycrystalline hetero junction thin films solar cells» and by the Ministry of Education and Science, Youth and Sport of Ukraine (Grant №. 01110U001151). The authors wish to thank Prof. Yu.P. Gnatenko and P.M. Bukivskij from the Institute of physics NAS of Ukraine for the PL measuring of some II-VI film samples.

7. References

[1] Georgobiani, A. (1974). Wide Band gap II-VI Semiconductors and Perspectives of Their Usage (in Russian), *Uspekhi Fizicheskikh Nauk*, Vol.113, No.1, pp. 129-155, ISSN 0042-1294.

[2] Pautrat, J. (1994). II–VI Semiconductor Microstructures: From Physics to Optoelectronics, *Journal de Physique III*, Vol.4, – pp. 2413-2425, ISSN 1155-4320.

[3] Takahashi, K., Yoshikawa, A. & Sandhu, A. (2007). *Wide Bandgap Semiconductors. Fundamental Properties and Modern Photonic and Electronic Devices*, Springer, ISBN, Berlin, Heidelberg, NewYork, Germany, USA.

[4] Owens, A. (2004). Compound Semiconductor Radiation Detectors, *Nuclear Instrumental Methods*, Vol. 531, pp. 18-37, ISSN 0168-9002.

[5] Grinev, B., Ryzhikov, V. & Seminozhenko, V. (2007). *Scintillation Detectors and Radiation Control Systems on Their Base (in Russian)*, Naukova Dumka, ISBN, Kyiv, Ukraine.

[6] Berchenko, N., Krevs, V. & Seredin, V. (1982). *Reference Tables (in Russian)*, Voenizdat, ISBN, Moskov, USSR.

[7] Ohring, M. (1992). *The Materials Science of Thin Films*, Academic Press, ISBN, NewYork, USA.

[8] Poortmans, J. & Arkhipov, V. (2006). *Thin Film Solar Cells: Fabrication, Characterization and Application*, John Wiley & Sons, Ltd. IMEC, ISBN, Leuven, Belgium.

[9] Panchekha, P. (2000). Structure and Technology Problems of II-VI Semiconductor Films, *Functional materials*, Vol.7, No.2, pp. 1-5, ISSN 1616-3028.

[10] Holt, D. & Yacobi, B. (2007). *Extended defects in semiconductors. Electronic properties, device effects and structures*, Cambridge University press, ISBN, New York, Melbourne, Madrid, Cape Town, USA, Australia, Spain, South Africa.

[11] Milns, A. (1977). *Impurities with Deep Levels in Semiconductors (in Russian)*, Mir, ISBN, Moskow, USSR.

[12] Shaskolskaya, M. (1984). *Crystallography: Studying Reference for Technical Universities (in Russian)*, Vysshaya Shkola, ISBN, Moskov, USSR.

[13] Korbutyak, D. & Melnychuk, S. (2000). *Cadmium Telluride: Impurity-Defect States and Detector Properties*, Ivan Fedorov, ISBN, Kyiv, Ukraine.

[14] Fochuk, P. (Experimental identification of the Point Defects/ P. Fochuk, O. Panchuk // CdTe and related Compounds: Physics, Defects, Hetero- and Nano-Structure, Crystal

growth, Surfaces and Applications. [R. Triboulet, P.Siffert]. Netherlands: Elsevier, 2010. – P. 292-362.

[15] Stokman, F. (1973) On the Classification of Traps and Recombination Centres, *Physica Status Solidi A*, Vol.20. – pp. 217-220, ISSN 1862-6319.

[16] Serdyuk, V., Chemeresyuk, G. & Terek, M. (1982). *Photoelectric Processes in Semiconductors*, Vyshcha Shkola, Main Edition, ISBN, Odessa, USSR.

[17] Meyer, B. & Stadler, W. (1996). Native Defect Identification in II-VI Materials, *Journal of Crystal Growth*, Vol.161, pp. 119-127, ISSN 0022-0248.

[18] Neumark, G. (1997). Defects in Wide Band Gap II-VI Crystals, *Material Science and Engineering A*, Vol.R21, No.1, - pp. 1-46, ISSN 0921-5093.

[19] Grundmann, M. (2010). *The Physics of Semiconductors. An Introduction Including Nanophysics and Applications*, Springer-Verlag, ISBN, Berlin, Heidelberg, Germany.

[20] Kosyak, V., Opanasyuk, A. & Panchal, J. (2011). Structural and Substructural Properties of the Zinc and Cadmium Chalcogenides (review), *Journal of Nano and Electronic Physics*, Vol.3, No.1, - pp. 274-301, ISSN 2077-6772.

[21] Pavlov, A. (1987). *Methods of Measuring the Semiconductor Materials Parameters (in Russian)*, Vysshaya Shkola, ISBN, Moskow, USSR.

[22] Vorobjev, Ju., Dobropolskiy, V. & Strikha, V. (1988). *Methods of Semiconductors Investigation*, Vyshcha Shkola, ISBN, Kyiv, USSR.

[23] Gorokhovatskij, Ju. & Bordovskij, G. (1991). *Thermoactivation Current Spectroscopy of High-ohmic Semiconductors and Dielectrics*, Nauka, ISBN, Moskow, USSR.

[24] Lampert, M. & Mark, P. (1973). *Injection Currents in Solid States*, Mir, ISBN, Moskow, USSR.

[25] Kao, K. (1984). *Electrons Transport in Solid States*. Vol.1, Mir, ISBN, Moskow, USSR.

[26] Lalitha, S., Sathyamoorthy, R. & Senthilarasu, S. (2004). Characterization of CdTe Thin Film-dependence of Structural and Optical Properties on Temperature and Thickness, *Solar Energy Materials & Solar Cells*, Vol.82, - pp. 187-199, ISSN 0927-0248.

[27] Ibrahim, A. (2006). DC Electrical Conduction of Zinc Telluride Thin Films, *Vacuum*, Vol.81, pp. 527–530, ISSN.

[28] Rose, A. (1955). Recombination Processes in Insulators and Semiconductors, *Physical Review*, Vol.97, pp. 322−323, ISSN 1943-2879.

[29] Rose A. (1955). Space–charge–limited Currents in Solids, *Physical Review*, Vol.97, pp. 1538–1544, ISSN 1943-2879

[30] Nespurek, S. & Sworakowsky, J. (1980). Evolution of Validity of Analytical Equations Describing Steady–state Space–charge–limited Current–voltage–characteristics, *Czechoslovak Journal of Physics*, Vol.B30, No.10, pp. 1148–1156, ISSN 0011-4626.

[31] Mark, P. & Helfrich, W. (1962). Space–charge–limited Currents in Organic Crystals, *Journal of Applied Physics*, Vol.33, pp. 205–215, ISSN 0021-8979.

[32] Thomas, J., Williams, J. & Turton L. (1968). Lattice Imperfections in Organic Solids. Part 3, 4, *Transactions of the Faraday Society*, Vol.64, pp. 2496−2504, ISSN 0014-7672.

[33] Hwang, W. & Kao, K. (1976). Studies of the Theory of Single and Double Injections in Solids with a Gaussian Trap Distribution, *Solid State Electronics*, Vol.19, pp. 1045–1047, ISSN 0038-1101.

[34] Nespurec, S. & Semejtec, P. (1972). Space Charge Limited Currents in Insulators With Gaussian Distribution of Traps, *Czechoslovak Journal of Physics*, Vol.B22, pp. 160–175, ISSN 0011-4626.

[35] Boncham, J. (1973). SCLC Theory for a Gaussian Trap Distribution, *Australian Journal of Chemistry*, Vol.26, pp. 927–939, ISSN 0004-9425.

[36] Simmons, J. & Tarn, M. (1973). Theory of Isothermal Currents and the Direct Determination of Trap Parameters in Semiconductors and Insulators Containing Arbitrary Trap Distributions, *Physical Review*, Vol.B7, pp. 3706-3713, ISSN 1943-2879.

[37] Pfister, J. (1974). Note of Interpretation of Space–charge–limited Currents With Traps, *Physica Status Solidi A*, Vol.24, No.1, pp. K15-K17, ISSN 1862-6319.

[38] Manfredotti, C., de Blasi, C. & Galassini, S. (1976). Analysis of SCLC Curves by a New Direct Method , *Physica Status Solidi A*, Vol.36, No.2, pp. 569-577, ISSN 1862-6319.

[39] Nespurek, S. & Sworakowski, J. (1978). A Differential Method of Analysis of Steady-state Space-charge-limited Currents: an Extension, *Physica Status Solidi A.*, Vol.49, pp. 149-152, ISSN 1862-6319.

[40] Nespurek, S. & Sworakowski, J. (1980). Use of Space–charge–limited Current Measurement to Determine of Properties of Energetic Distributions of Bulk Traps, *Journal of Applied Physics*, Vol.51, No.4, pp. 2098-2102, ISSN 0021-8979.

[41] Nespurek, S. & Sworakowski, J. (1990). Spectroscopy of Local States in Molecular Materials Using Space-charge-limited Currents, *Radiation Physics and Chemistry*, Vol.36, No.1, pp. 3-12, ISSN 0969-806X.

[42] Nespurec, S., Obrda, J. & Sworakowsky, J. (1978). Study of Traps for Current Carriers in Organic Solids N,N' – Diphenyl–p–Phenylenediamine, *Physica Status Solidi A*, Vol.46, No.1, pp. 273–280, ISSN 1862-6319.

[43] Lyubchak, V., Opanasyuk, A. & Tyrkusova, N. (1999). Injection Spectroscopy Method for investigation of Deep Centers in Cadmium Telluride Films (in Ukrainian), *Ukrayinskyi Fizychnyi Zhurnal*, Vol.44, No.6, pp. 741-747, ISSN 0503-1265.

[44] Opanasyuk, A., Tyrkusova, N. & Protsenko, I. (2000). Some Features of the Distributions Deep States Reconstruction by Injection Spectroscopy Method (in Ukrainian), *Zhurnal Fizychnykh Doslidzhen*, Vol.4, No.2, pp. 208-215, ISSN 1027-4642.

[45] Opanasyuk, A., Opanasyuk, N. & Tyrkusova, N. (2003), High-temperature Injection Spectroscopy of Deep Traps in CdTe Polycrystalline Films, *Semiconductor Physics, Quantum Electronics & Optoelectronics*, Vol.6, No.4, pp. 444-449, ISSN 1560-8034.

[46] Tikhonov, A. & Jagola, A. (1990). *Numerical Nethods of Non-correct Problems Solution*, Nauka, ISBN, Moskow, USSR..

[47] Zavjalov, Ju., Kvasov, B. & Miroshnichenko, V. (1980). *Spline-functions Methods*, Nauka, ISBN, Moskow, USSR..

[48] Turkusova, N. (2002). *Injection Spectroscopy of Deep Trap Levels in Cadmium Telluride Films, PhD Thesis (in Ukrainian)*, Sumy State University, Sumy, Ukraine.

[49] Kosyak, V., Opanasyuk, A. & Bukivskij, P. (2010). Study of the Structural and Photoluminescence Properties of CdTe Polycrystalline Films Deposited by Closed Space Vacuum Sublimation, *Journal of Crystal Growth*, Vol.312, pp. 1726-1730, ISSN 0022-0248.

[50] Kurbatov, D., Khlyap, H. & Opanasyuk, A. (2009). Substrate–temperature Effect on the Microstructural and Optical Properties of ZnS Films Obtained by Close–spaced Vacuum Sublimation, *Physica Status Solidi A*, Vol.206, No.7, pp. 1549-1557, ISSN 1862-6319.

[51] (1988). *Selected Powder Diffraction Data for Education Straining (Search Manual and Data Cards)*, International Centre for Diffraction Data, USA.

[52] Zjuganov, A. & Svechnikov, S. (1981). *Injection-contact Effects in Semiconductors (in Russian)*, Naukova Dumka, ISBN, Kiev, USSR.

[53] Zjuganov, A., Smertenko, P. & Shulga, E. (1979). Generalized Method of Determination the Volume and Contact Semiconductor Parameters by Current-voltage characteristic (in Russian), *Poluprovodnikovaja Tekhnika i Elektronika*, Vol.29, pp.48-54, ISSN.

[54] Kurbatov, D., Kolesnyk, M. & Opanasyuk, A. (2009). The Substructural and Optical Characteristics of ZnTe Thin Films, *Semiconductor Physics, Quantum Electronics & Optoelectronics*, Vol.12, No.1, pp. 35-40, ISSN 1560-8034.

[55] Kurbatov, D., Denisenko, V. & Opanasyuk, A. (2008). Investigations of Surface Morphology and Chemical Composition Ag/ZnS/Glassceramic Thin Films Structure, *Semiconductor Physics, Quantum Electronics & Optoelectronics*, Vol.11, No.4, pp. 252-256, ISSN 1560-8034.

[56] Balogh, A., Duvanov, D. & Kurbatov, D. (2008). Rutherford Backscattering and X–ray Diffraction Analysis of Ag/ZnS/Glass Multilayer System, *Photoelectronics*, Vol.B.17, pp.140-143, ISSN 0235-2435.

[57] Wei, S. & Znang, S. (2002). Chemical Trends of Defect Formation and Doping Limit in II–VI Semiconductor: The Case of CdTe, *Physical Review B*, Vol.66, pp. 1-10, ISSN 0163-1829.

[58] Wei, S. & Zhang, S. (2002). First-Principles Study of Doping Limits of CdTe, *Physica Status Solidi B*, Vol.229, No. 1, pp. 305–310, ISSN 0370-1972.

[59] Soundararajan, R., lynn, K. & Awadallah, S. (2006). Study of Defect Levels in CdTe Using Thermoelectric Effect Spectroscopy, *Journal of Electronic Materials*, Vol.35, No.6, pp., ISSN 0361-5235.

[60] Berding, M. (1999). Native Defects in CdTe, *Physical Review B*, Vol.60, No.12, pp. 8943-8950, ISSN 0163-1829.

[61] Berding M. (1999). Annealing Conditions for Intrinsic CdTe, *Applied Physics Letters*, Vol.74, No.4, pp. 552-554, ISSN 0003-6951.

[62] X.Mathew. Photo-induced current transient spectroscopic studu of the traps in *CdTe// Solar Energy Materials & Solar Cells*. 76, pp. 225-242 (2003).

[63] Mahalingam, T., John, V. & Ravi, G. (2002). Microstructural Characterization of Electrosynthesized ZnTe Thin films, *Crystal Research and Technology*, Vol.37, No.4, pp. 329–339, ISSN 1521-4079.

[64] Ibrahim, A, El-Sayed, N. & Kaid, M. (2004). Structural and Electrical Properties of Evaporated ZnTe Thin Films, *Vacuum*, Vol.75, pp. 189–194, ISSN 0042-207X.

[65] Dean, P., Venghaus, H. & Pfister, J. (1978). The Nature of the Predominant Acceptors in High Quality Zinc Telluride, *Journal of Luminescence*, Vol.16, pp. 363-394, ISSN 0022-2313.

[66] Bhunia, S., Pal, D. & Bose, N. (1998). Photoluminescence and Photoconductivity in Hydrogen–passivated ZnTe, *Semiconductor Science and Technology*, Vol.13, pp. 1434-1438, ISSN 0268-1242.

[67] Sadofjev, Ju. & Gorshkov, M. (2002). Deep Level Spectra in ZnTe:Cr^{2+} Layers Obtained by Epitaxy from Molecular Beams (in Russian), *Fizika I Tekhnika Poluprovodnikov*, Vol.36, No.5, pp. 525-527, ISSN 0015-3222.

[68] Korbutyak, D., Vakhnyak, N. & Tsutsura, D. (2007). Investigation of Photoluminescence and Electroconductivity of ZnTe Grown in Hydrogen Atmosphere, *Ukrainian Journal of Physics*, Vol.52, No.4, pp. 378-381, ISSN 2071-0186.

[69] Feng, L., Mao, D. & Tang, J. (1996). The Structural, Optical, and Electrical Properties of Vacuum Evaporated Cu-doped ZnTe Polycrystalline Thin Films, *Journal of Electronic Materials*, Vol. 25, pp. 1422–1427, ISSN 0361-5235.

[70] Atakova, M., Ramazanov, P. &Salman, E. (1973). Local Levels in a Zinc Sulfide Film, *Izvestiya Vysshikh Uchebnykh Zavedenii, Fizika*, Vol.10, pp. 95 – 98, ISSN 0021-3411.

[71] Venkata Subbaiah Y., Prathap, P. & Ramakrishna Reddy, K. (2006). Structural, Electrical and Optical Properties of ZnS Films Deposited by Close-spaced Evaporation, *Applied Surface Science*, Vol. 253, pp. 2409 – 2415, ISSN 0169-4332.

[72] Venkata Subbaiah, Y., Prathap, P. & Ramakrishna Reddy, K. (2008). Thickness Effect on the Microstructure, Morphology and Optoelectronic Properties of ZnS Films, *Journal of Physics: Condensed Materials*, Vol.20, pp. 035205 – 035215, ISSN 0953-8984.

[73] Turan, E., Zor, M. & Aybek, A. (2007). Thermally Stimulated Currents in ZnS Sandwich Structure Deposited by Spray Pyrolysis, *Physica B: Condensed Materials*, Vol.395, No.1, pp. 57 – 64, ISSN 0921-4526.

[74] Abbas, J., Mehta, C. & Saini, G. (2007). Preparation and Characterization of n-ZnS and its Self-assembled Thin Film, *Digest Journal of Nanomaterials and Biostructures*, Vol.2, No.3, pp. 271 – 276, ISSN 1842-3582.

[75] Kvit, A., Medvedev, S. & Klevkov, Ju. (1998). Optical Spectroscopy of Deep States in ZnTe (in Russian), *Fizika I Tekhnika Poluprovodnikov*, Vol.40, No.6, pp. 1010-1017, ISSN 0015-3222.

[76] Bagaev, V., Zajtsev, V. & Klevkov, Ju. (2003). Influence of Annealing in Vapors and in Liquid Zn on ZnTe High-frequency Polycrustalline Photoluminescence (in Russian), *Fizika I Tekhnika Poluprovodnikov*, Vol.37, No.3, pp. 299-303, ISSN 0015-3222.

[77] Klevkov, Ju., Martovitskij, V., Bagaev, V. (2006). Morphology, Twins Formation and Photoluminescence of ZnTe crystals Grown by Chemical Synthesis of Components From Vapor Phase (in Russian), *Fizika I Tekhnika Poluprovodnikov*, Vol.40, No.2, pp. 153-159, ISSN 0015-3222.

[78] Klevkov, Ju., Kolosov, S. & Krivobokov, V. (2008). Electrical Properties, Photoconductivity and Photoluminescence of large-crystalline p-CdTe (in Russian), *Fizika I Tekhnika Poluprovodnikov*, Vol.42, No.11, pp. 1291-1296, ISSN 0015-3222.

[79] Bagaev, V., Klevkov, Ju. & Krivobokov, V. (2008). Photoluminescence Spectrum Change Near the Twins Boundaries in ZnTe Crystals Obtained by High-speed Crystallization (in Russian), *Fizika Tverdogo Tela*, V.50, No.5, pp. 774-780, ISSN 0042-1294.

[80] Makhniy V. & Gryvun, V. (2006). Diffusion ZnTe:Sn Layers with Electron Conductivity (in Russian), *Fizika I Tekhnika Poluprovodnikov*, Vol.40, No.7, pp. 794-795, ISSN 0015-3222.

[81] Tsutsura, D., Korbutyak, O. & Pihur, O. (2007). On Interaction of Hydrogen Atoms With Complex Defects in CdTe and ZnTe, *Ukrainian Journal of Physics*, Vol.52, No.12, pp. 1165-1169, ISSN 2071-0186.

[82] Taguchi, T, Fujita, S. & Inushi, Y. (1978). Growth of High-purity ZnTe Single Crystals by the Sublimation Travelling Heater Method, *Journal of Crystal Growth*, Vol.45, pp. 204-213, ISSN 0022-0248.

[83] Dean, P. (1979). Copper, the Dominant Acceptor in Refined, Undoped Zinc Telluride, *Luminescence*, Vol.21, pp. 75-83, ISSN 1522-7243.

[84] Garcia, J., Remon, A. & Munoz, V. (2000). Annealing-induced Changes in the Electronic and Structural Properties of ZnTe Substrates, *Journal of Materials Research*, Vol.15, No.7, pp. 1612-1616, ISSN 0884-2914.

[85] Uen, W., Chou, S. & Shin, H. (2004), Characterizations of ZnTe Bulks Grown by Temperature Gradient Solution Growth, *Materials Science & Engineering A*, Vol.B106, pp. 27-32, ISSN 0921-5093.

[86] Yoshino, K, Yoneta, V. & Onmori, K. (2004). Annealing Effects of High–quality ZnTe Substrate, *Journal of Electronic Materials*, Vol.33, No.6, pp. 579-582, ISSN 0361-5235.

[87] Yoshino, K., Kakeno, T. & Yoneta, M. (2005). Annealing Effects of High–quality ZnTe Substrate, *Journal of Material Science: Materials in Electronics*, Vol.16, pp. 445-448, ISSN.0957-4522.

[88] Ichiba, A., Ueno, J. & Ogura, K. (2006). Growth and Optical Property Characterizations of ZnTe:(Al, N) layers Using Two Co–doping Techniques, *Physica Status Solidi C*, Vol.3, No.4, pp. 789-792, ISSN 1610-1642.

[89] Bose, D. & Bhunia, S. (2005). High Resistivity In–doped ZnTe: Electrical and Optical Properties, *Bulletin of Material Science*, Vol.28, No.7, pp. 647-650, ISSN 0250-4707.

[90] Grill, R., Franc, J, & Turkevych, I. (2002). Defect-induced Optical Transitions in CdTe and $Cd_{0.96}Zn_{0.04}Te$ *Semiconductor Science and Technology*, Vol.17, pp. 1282-1287, ISSN 0268-1242.

[91] Ushakov, V. & Klevkov, Ju. (2007). Microphotoluminescence of Undoped Cadmium Telluride Obtained by Non-equilibrium Method of Direct Synthesis in Components Fluctuation (in Russian), *Fizika I Tekhnika Poluprovodnikov*, Vol.41, No.2, pp. 140-143, ISSN 0015-3222.

[92] Babentsov, V., Corregidor, V. & Castano, J. (2001). Compensation of CdTe by Doping With Gallium, *Crystal Research and Technology*, Vol.36, No.6, pp. 535-542, ISSN 1521-4079.

[93] Tarbayev, G. & Shepelskij. (2006). Two Series of "Dislocation" Photoluminescence Lines in Cadmium Telluride Crystals (in Russian), *Fizika I Tekhnika Poluprovodnikov*, Vol.40, No.10, pp. 1175-1180, ISSN 0015-3222.

[94] Aguilar-Hernandez, J., Contreras-Puente, J. & Vidal-Larramendi J. (2003). Influence of the Growth Conditions on the Photoluminescence Spectrum of CdTe Polycrystalline Films Deposited by the Close Space Vapor Transport Technique, *Thin Solid Films*, Vol.426, pp. 132-134, ISSN 0040-6090.

[95] Aguilar-Hernandez, J., Cardenas-Garcia, M. & Contreras-Puente, G. (2003). Analysis of the 1,55 eV PL Band of CdTe Polycrystalline Films, *Material Science & Engineering A*, Vol.B102, pp. 203-206, ISSN 0921-5093.

[96] Armani, N., Ferrari, C. & Salviati, G. (2002). Defect-induced Luminescence in High–resistivity High–purity Undoped CdTe Crystals, *Journal of Physics: Condensed Matterials*, Vol.14, pp. 13203-13209, ISSN 0953-8984.

[97] Gheluwe, J., Versluys, J. & Poelman, D. (2005). Photoluminescence Study of Polycrystalline CdS/CdTe Thin Film Solar Cells, *Thin Solid Films*,Vol.480–481, pp. 264-268, ISSN 0040-6090.

[98] Okamoto, T., Yamada, A. & Konagai, M. (2001). Optical and Electrical Characterizations of Highly Efficient CdTe Thin Film Solar Cells, *Thin Solid Films*, Vol.387, pp. 6-10, ISSN 0040-6090.

[99] Holliday, D., Potter, M. & Boyle, D. (2001). Photoluminescence Characterisation of Ion Implanted CdTe, *Material Research Society, Symposium Proceedings*, Vol.668, pp. H1.8.1-H1.8.6, ISSN 0272-9172.

[100] Palosz, W., Grasza, K. & Boyd, P. (2003). Photoluminescence of CdTe Crystals Grown Physical Vapor Transport, *Journal of Electronic Materials*, Vol.32, No.7, pp.747-751, ISSN 0361-5235.

[101] Corregidor, V., Saucedo, E. & Fornaro, L. (2002). Defects in CdTe Polycrystalline Films Grown by Physical Vapour Deposition, *Materials Science and Engineering A*, Vol.B 91–92, pp. 525-528, ISSN 0921-5093.

[102] Soundararajan, R., Lynn, K. & Awadallah, S. (2006). Study of Defect Levels in CdTe Using Thermoelectric Effect Spectroscopy, *Journal of Electronic Materials*, Vol.35, No.6, pp. 1333-1340, ISSN 0361-5235.

[103] Babentsov, V. & Tarbaev. (1998). Photoluminescence of Re-crystallized by Nano-second Laser Irradiation Cadmium Telluride (in Russian), *Fizika I Tekhnika Poluprovodnikov*, Vol.32, No.1, pp. 32-35, ISSN 0015-3222.

[104] Morozova, N. & Kuznetsov, V. (1987). *Zinc Sulfide. Obtaining and Optical Properties (in Russian)*, Nauka, ISBN, Moskow, USSR.

[105] Sahraei, R., Aval, J. & Goudarzi, A. (2008). Compositional, Structural and Optical Study of Nanocrystalline ZnS Thin Films Prepared by a New Chemical Bath Deposition Route, *Journal of Alloys and Compounds*, Vol.466, pp. 488-492, ISSN 0925-8388.

[106] Prathap, P. Revathi, N. & Venkata Subbaiah, Y. (2008). Thickness Effect on the Microstructure, Morphology and Optoelectronic Properties of ZnS Films, *Journal of Physics: Condensed Materials*, Vol.20, pp. 035205-035215, ISSN 0953-8984.

[107] Prathap, P. & Venkata Subbaiah, Y. (2007). Influence of Growth Rate on Microstructure and Optoelectronic Behaviour of ZnS Films, *Journal of Physics D: Applied Physics*, Vol. 40, pp. 5275-5282, ISSN 0022-3727.

[108] Lee, H. & Lee, S. (2007). Deposition and Optical Properties of Nanocrystalline ZnS Thin Films by a Chemical Method , *Current Applied Physics*, Vol.7, pp. 193-197, ISSN 1567-1739.

[109] Kroger, F. (1964). *The chemistry of Imperfect Crystals*, North-holland publishing company, ISBN, Amsterdam, Netherlands.

[110] Grill, R. & Zappetini, A. (2004). Point Defects and Diffusion in Cadmium Telluride, *Progress in Crystal Growth and Characterization of Materials*, Vol.48, pp .209-244, ISSN 0146-3535.

[111] Kosyak, V. & Opanasyuk, A. (2005). Point Defects Encemble in CdTe Single Crystals in the Case of full Equilibrium and Quenching (in Ukrainian), *Fizyka I Himiya Tverdych Til*, Vol.6, No.3, pp. 461-471, ISSN 1729-4428.

[112] Kosyak, V., Opanasyuk, A. & Protsenko, I. (2005). Ensemble of Point Defects in Single Crystals and Films in the Case of Full Equilibrium and Quenching, *Functional Materials*, Vol.12, No.4, pp. 797-806, ISSN 1616-3028.

[113] Kosyak, V. & Opanasyuk, A. (2007). Calculation of Fermi Level Location and Point Defect Ensemble in CdTe Single Crystal and Thin Films, *Semiconductor Physics, Quantum Electronics & Optoelectronics*, Vol.10, No.3, pp. 95-102, ISSN 1560-8034.

[114] Kolesnik, M., Kosyak, V. & Opanasyuk, A. (2007). Calculation of Point Defects Ensemble in CdTe Films Considering Transport Phenomenon in Gas Phase, *Radiation Measurements*, Vol.42, No.4–5, pp. 855-858, ISSN 1350-4487.

[115] Kosyak, V., Kolesnik, M. & Opanasyuk, A. (2008). Point Defect Structure in CdTe and ZnTe Thin Films, *Journal of Material Science: Materials in Electronics*, Vol.19, No.1, pp. S375-S381, ISSN 0957-4522.

[116] Kurbatov, D, Kosyak, V., Opanasyuk, A. (2009). Native Point Defects in ZnS Films, *Physica. B. Condensed Materials*, Vol.404, No.23–24, pp. 5002-5005, ISSN 0921-4526.

Chapter 8

Mathematical Methods to Analyze Spectroscopic Data – New Applications

E.S. Estracanholli, G. Nicolodelli, S. Pratavieira, C. Kurachi and V.S. Bagnato

Additional information is available at the end of the chapter

1. Introduction

Absorption and fluorescence spectroscopies in the visible and/or infrared spectrum are good options when a fast and objective analysis is required. Spectroscopy is based on light-matter interactions. This interaction occurs in different ways, and each molecule or an ensemble of molecules will show a distinctive response. The vibrational spectroscopy provides a fingerprint of the vibrational levels of a molecule usually at mid-infrared (MIR) radiation (400-4000cm^{-1}). The optical spectroscopy uses the ultraviolet-visible (UV-VIS) region (200-1000nm) of the electromagnetic spectrum and interrogates the electronic levels of a molecule. The instrumentation used to generate and detect this radiation is less complex and cheaper compared to other spectroscopy techniques, such as nuclear magnetic resonance, X-rays, etc. An absorption spectrum is obtained by irradiating a sample and measuring the light which is transformed into other forms of energy, e.g. molecular vibration (heat). A fluorescence spectrum is obtained only from fluorescent molecules, those that absorb and then emit radiation, acquiring the intensity of light emitted as a function of the wavelength. These spectra are characteristic for each molecule, because each one has different electronic levels and vibrational modes. These levels and modes are also influenced by the solvent of the molecule.

Vibrational spectroscopy applied to the mid portion of the infrared spectrum provides the basis to develop several of the most powerful methods of qualitative and quantitative chemical analysis. Some of the advantages to use this technique are: the information is collected on a molecular level, almost any chemical group has IR bands, it is very environment-sensitive.[1-3]

Optical fluorescence spectroscopy is highly sensitive and can provide different information about the molecules and the molecular processes such as the molecular interaction with the environment, the molecular bonding and concentration.[3, 4]

Spectroscopic techniques in this range of the electromagnetic spectrum have shown applications in different areas, from analytical chemistry to the diagnosis of some types of cancer, detection of citrus diseases and of dental caries, with a high sensitivity and good specificity rates. This is possible because the analyzed systems are composed of different types and concentrations of molecules. Thus, the spectrum of samples obtained under different conditions will also be different. It is therefore possible to identify and also quantify different compounds. However, the spectral variation can be characterized and correlated only with difficulty. This is mainly due to the fact that other phenomena, such as scattering and/or absorption, happen with the emitted light. In some cases, there may be other molecules in the sample presenting absorption bands that overlap in the same spectral region of the compound of interest, this mainly happens for absorption spectroscopy. In other cases, as in the case of fluorescence spectroscopy, the excitation and emitted light can be absorbed by other molecules making the signal too weak to be detected. A solution to this problem may be statistical procedures applied where the spectral information is correlated with any parameter of interest. [3-5]

A new application of a statistical method to process multi-layer spectroscopy information will be presented in this chapter. A brief review of the mathematical methods to analyze these spectroscopy data will be shown here, followed by two distinct examples. The first example is UV-VIS fluorescence spectroscopy, applied to detect the postmortem interval (PMI) in an animal model. The spectroscopy and statistical methods of analysis presented can be extended to other samples, like food and beverage. Here, a MIR absorption spectroscopy of liquid samples will be presented to detect and quantify certain compounds during the production of beer. Another system to measure liquid samples, which consists of a sample holder, will also be presented. This system offers a cheaper technique with a better signal compared to techniques used to analyze liquid samples in the MIR region.

2. Pattern recognition in the complex spectral database – Example of fluorescence spectroscopy used in forensic medicine

One of the limitations of conventional methods to determine the Postmortem Interval (PMI) of an individual is the fact that the measurements cannot be performed in real time and *in situ*. Several factors, environmental and body conditions influence the tissue decomposition and the time evolution, resulting in a poor resolution. Considering this limitation to determine the PMI, a possible solution is a new more objective method based on a tissue characterization of the degradation phases through optical information using fluorescence spectroscopy. If proven sensitive enough, this method shows a main advantage over conventional methods: less inter-variance and quantitative tissue information. These characteristics are relevant because they are less influenced by individual skills.[6]

During the decomposition process a wide variety of organic materials are consumed by natural micro-organisms and other unknown compounds produced by them. Using an objective method based on the tissue characterization of the stages of degradation by means of optical information using ultraviolet-visible fluorescence spectroscopy, followed by a statistical method based on PCA (Principal Component Analysis) made it possible to identify well features with time progression. The characteristic pattern of time evolution presented a high correlation coefficient, indicating that the chosen pattern presented a direct linear relationship with the time evolution. The results show the potential of fluorescence spectroscopy to determine the PMI, with at least a similar resolution compared to conventional methods.[6]

Another attractive feature of optical technologies is the fact that in situ information is achieved through a noninvasive and nondestructive interrogation with a fast response. Conventional laboratory techniques to determine PMI are time consuming and also require the cadaver removal from the location where it was found to a forensic lab facility. This operation already introduces additional changes to the analysis.

Fluorescence spectroscopy has been presented as a sensitive technique to biochemical and structural changes of tissues. The investigation of biological tissues is quite complex. Photons interact with biomolecules in several ways, and depending on the type of the interactions, they can be classified into three groups. The absorbers are the biomolecules that absorb photon energy. The fluorophores are biomolecules that absorb and emit fluorescent light. The scatterers are biomolecules that do not absorb the photons but change their direction. Several endogenous cromophores contribute and modify the final tissue spectrum. Distinct fluorophores emit light but the collected spectrum will be modified depending on the presence of absorbers and scatterers in the microenvironment on the path of the emitted photons and the probe interrogator. Taking into account all these light interactions that occur within the biological tissues, it is important to keep in mind that a tissue fluorescence spectrum is a result of the combination of all these processes occurring in the pathway between excitation and collection: excitation absorption and scattering, fluorescence emission, and fluorescence absorption and scattering.[5, 7]

Tissue changes begin to take place in the cadaver as soon as there is cessation of life. Optical characteristics change, and these changes may be detected using fluorescence spectroscopy. With the cessation of the metabolic reactions, tissue modifications are induced by several distinct factors, e.g. lack of oxygen and adenosine triphosphate, and intestinal microorganism proliferation. In this type of analyses, we first aimed to establish a proof of concept that fluorescence spectral variations for distinct PMIs are higher than the variance observed within each PMI. If the results are positive, a spectral time behavior can then be determined, i.e. fluorescence spectral changes identifying each PMI. This proposed method can be used to determine an unknown PMI based on a comparative analysis of a spectral database pattern. There is a potential correlation of the tissue fluorescence changes and the PMI, even though the same limitations concerning the time course variability of the cadaveric phenomena are also present, the optical spectra information can provide a more objective estimation.

Taking into account the resolution limitation to determine PMI in biological tissues, where the degradation process is non-homogenous and influenced by environment and cadaver intrinsic factors, the optical techniques may show a better PMI prediction when compared to current techniques.

The process using the principal component analysis is necessary to change the space analysis. For any type of spectroscopy, where space is determined by an analysis of light intensity in wavelengths of absorption or fluorescence, a change of base can be accomplished, where the variables become the variance of the dataset. We can explore this idea mathematically using a practical example. The set of spectra shown in figure 1 is a typical result of multiple samples.

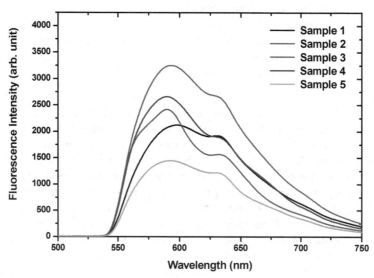

Figure 1. Fluorescence spectra of different samples.

In this case, we are dealing with intensities as a function of the wavelength, presented in a graphic form. These same curves can be represented in a matrix, as shown in table 1. In this form, each row corresponds to a single measurement, i.e., spectrum of a sample, and each column is the value of the wavelength considered, which makes up this spectrum. Thus, each array element is the intensity measured at a wavelength specific to a spectrum.

Wavelength (nm)	540	541	542	543	...	...	749	750
Intensity Sample1	27.5	30.6	33.9	37.7	...	...	199.9	198.4
Intensity Sample2	25.5	26.7	29.6	35.9	...	...	180.5	186.3
...	...	...	...	...	...	...	...	...
Intensity SampleN	24.1	25.3	27.9	30.2	...	...	176.8	180.6

Table 1. Fluorescence spectra in a matrix.

The next procedure to be performed after inserting the data set into this matrix representation is the centralization of the data around their average value. By fixing one column (wavelength) at a time we can calculate the average value for all lines (samples). Table 2 shows the mean values obtained.

Wavelength (nm)	540	541	542	543	...	...	749	750
Intensity Media	25.7	27.53	30.47	34.6	...	...	185.73	188.43

Table 2. Matrix of mean values for each wavelength.

In the next step, each of the intensity values of each sample should be subtracted from this average value in the respective wavelength. The results for our example are shown in table 3.

Wavelength (nm)	540	541	542	543	...	...	749	750
Intensity Sample1	1.8	3.07	3.43	3.1	...	...	14.17	9.97
Intensity Sample2	-0.2	-0.83	-0.87	1.3	...	...	-5.23	-2.13
...	...	...	...	...	...	...	...	...
Intensity SampleN	-1.6	-2.23	-2.57	-4.4	...	...	-8.93	-7.83

Table 3. Normalized fluorescence spectra.

It is important to note that this procedure resulted in a better match between the variables. The first values had higher intensities (approximately 8 times) than the wavelengths around 750nm in relation to values around 540 nm. If this normalization of the data had not been performed, the outcome would have had a greater influence for the longer wavelengths, as if they possessed some kind of "privilege", which would not be correct from the standpoint that all the measured variables are also important.

Since each element of the initial data array is represented by an element q_{ij}, we can consider this procedure performed using the equation 1:

$$X_{ij} = q_{ij} - \overline{q}_j \tag{1}$$

Where: x_{ij} is the element of our new data matrix; q_{ij} is the array element data corresponding to the i-th measurement variable j; $\overline{q}_j$ is the mean value of the variable j;

As the data were previously normalized, i.e. centered on their mean values, we proceed with the construction of the correlation matrix, where we obtain information about a dataset that indicates how the variables are correlated. This is possible by calculating the product of the transposed data matrix by itself. In mathematical terms, if x is our new array of standardized data x_{ij} composed of elements, then the correlation matrix R formed by these correlation coefficients is given by:

$$R = X^T \cdot X \tag{2}$$

A matrix whose elements are given by:

$$r_{jj'} = \sum_{i=1}^{n} x_{ij} x_{ij'} = \sum_{i=1}^{n} \frac{(q_{ij} - \overline{q}_j) \cdot (q_{ij'} - \overline{q}_{j'})}{\sigma_j \cdot \sigma_{j'}} \tag{3}$$

The r_{jj} value is a standardized covariance between -1 and 1. It should be noted that the matrix is Hermitian (symmetric in the case of real variables, which is our case). We can also confirm that the elements along the main diagonal of the correlation matrix (elements where $j = j'$) correspond to the variance of the variable q_j. As noted earlier, the correlation matrix R is Hermitian and therefore its eigenvalues are real and positive and its eigenvectors are orthogonal. For this procedure, it is important to note that the values of the wavelengths themselves are not taken into account in the mathematical calculation. The data selected for the next step are the rows and columns highlighted in table 3.

After calculating all the elements of the correlation matrix, the diagonalization is necessary. The diagonalization process provides two sets of data. The first are the eigenvectors: vectors which constitute a new base having the direction and sense in which the initial data set has more tendencies to vary, i.e. the maximization of the variance. The second sets of data are the eigenvalues, which provide the weight information, i.e. the relevance of each of the directions of the eigenvectors.

The eigenvalues are represented by the matrix K and the eigenvectors by the matrix V:

$$K = \begin{bmatrix} \lambda_1 & 0 & \cdots & 0 \\ 0 & \lambda_2 & \cdots & 0 \\ \cdots & \cdots & \cdots & 0 \\ 0 & 0 & \cdots & \lambda_n \end{bmatrix} \tag{4}$$

As the diagonalized matrix was a correlation matrix, we assume that this new base formed by the set of eigenvectors of each R represents a percentage of the total variance; and the information contained in each eigenvector are unique and exclusive, since they are mutually orthogonal. Each λ_i in the K matrix represent the weight of the specific eigenvector. Through these eigenvalues we can determine which of the principal components explains the greatest amount of data. For the simple relationship between the value of each eigenvalue divided by the sum of all eigenvalues, i.e. $\lambda_i / \sum \lambda_i$, we can determine the weight (or representation) of each eigenvector.

Once we determined the basis that maximizes the variance of the data set, which should be done by "projecting" the initial data matrix in this new basis through the product between the eigenvectors and the matrix of normalized data, we obtain:

$$S = V \cdot Q \tag{5}$$

This data set designed in the new base (matrix S) is known as *Score*. The transformation of the basic matrix data S into the matrix Q by the base which maximizes the variances is known as Karhunen-Loève transformation.

The matrix of scores representing the data in our new base is expressed in such a way that each column represents the projection of the initial data into one of the eigenvectors, or in other terms, in either direction variance. Each line of the S matrix still represents a measure, or spectrum, as shown in table 4.

	PC1	PC2	PC3	PC4	...	...	PC N-1	PCN
Sample1	0.6	0.4	0.7	1.1	...	...	2.0	1.7
Sample2	0.8	1.2	1.4	2.2	...	...	1.7	0.6
...	...	...	...	...	...	...	...	...
SampleN	0.2	2.1	0.1	0.3	...	...	0.9	0.5

Table 4. Presentation of the available data with the new basis

Now, instead of analyzing the data obtained on the basis of the variables that were defined as the value of the intensity at each wavelength, these are considered in the space of the variances of these values. This change of base allows a significant reduction of the information in which the data are analyzed.

Spectroscopy experiments measure intensity values in hundreds or thousands of wavelengths, providing up to hundreds or thousands of variables to be analyzed. Depending on the experiment, when calculating the principal components of this system we represent around 90% of the information system, i.e. the spectra obtained in only two components of this new base. The graph of two major principal components (PC1 versus PC2) provides a much better view to then analyze these data rather than the hundreds of dimensions we had obtained before. In other words, we now work with a significantly reduced number of variables, wavelengths, with no loss of information. On this new basis, each sample, which was previously represented graphically by a curve with hundreds of points, shall be represented by a single point only. This significant reduction and simplification makes it much easier to detect spectral patterns.

We will now go back to the example of determining the postmortem interval - the set of measures shown in figure 1. Each sample is a fluorescence spectrum from different postmortem intervals. We apply the above procedure and obtain results on this new basis. For this case, the first two principal components show a representation greater than 91%. These results are shown in figure 2, where each point represents a spectrum.

Comparing data of figure 2 to those of figure 1 a temporal evolution of standard measurements becomes obvious. Based on this analysis, each region of space PC1 x PC2 is characterized by a postmortem interval. So in terms of a practical application, if we have a spectrum obtained from an unknown postmortem interval, we just design it based on this new basis and thus allow matching of the region of space to this spectrum, represented by a new point. Thus, this procedure can be used to determine a postmortem interval using fluorescence spectroscopy for situations where this value is unknown.

The present methodology and results shown by Estracanholli et al.[6, 8] demonstrated the use of fluorescence tissue spectroscopy to determine PMI as a valuable tool in forensic medicine. Two approaches were employed to associate the spectral changes with the time evolution of tissue modification. First, direct spectral changes were computed using inter-spectra analysis, allowing establishing a pattern of the sample distribution with a time evolution. Second, the use of a statistical method based on PCA helped to identify features over time. In both cases, the characteristic time evolution pattern presented a high correlation coefficient, indicating that the chosen pattern presented a direct linear relationship with time.

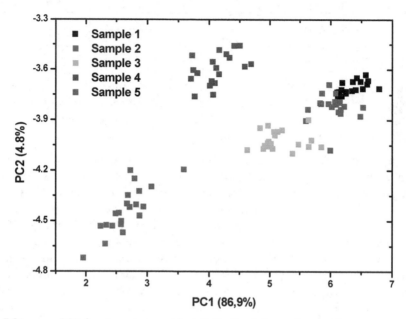

Figure 2. PC1 versus PC2 showing a more evident distinction between the samples.

However, other cases of application of spectroscopic techniques require a more robust processing. One such case is the other example cited above, where the goal is to quantify various compounds in a complex sample containing various interferences, and this most often occurs in the regions of overlapping absorption (or transmission and emission) of compounds of interest. For these cases, the application of artificial neural networks is a powerful solution.

3. Quantification of a composite with overlapping bands - Using vibrational spectroscopy in a beer sample

Beer brewing is a relatively long and complex biotechnological process, which can generate a range of products with distinct quality and organoleptic characteristics, all of which may be relevant to determine the type of product that should be made. Failures during important steps such as saccharification and fermentation can lead to major financial losses, i.e. to a loss of a whole batch of beer. Currently, analyses of the physic-chemical processes are carried out offline using traditional tests which do not provide any immediate response, e.g. HPLC (High Performance Liquid Chromatography). In the case of micro- breweries, which currently increase, some of these tests cannot be performed at all, due to the prohibitive cost of these tests. Therefore, many breweries do not have a possibility to identify errors during the production and to take corrective actions early-on. Today, problems are detected only at a later stage, towards the end of the brewing process. Currently, most systems used in the breweries consume time and can potentially compromise the quality of a whole batch. A solution to this problem is a new method consisting of a system to monitor in real time (online) the saccharification and fermentation steps of the wort. The amounts of alpha-amylase and beta-amylase in the grain are correlated with the time required to convert all grain starch into sugars. [3, 9-11]

The brewing-process is based on traditional recipes, a defined period of time and temperature. The amount of different types of enzymes in the grain when the wort is produced is, however, not known, since this amount depends on many factors, e.g. storage conditions, temperature, humidity, transport. Due to these factors, the saccharification step could be stopped, which would mean that a significant amount of starch would remain in the wort, and therefore the procedure would result in poor wort. Or, it is also possible that all starch may have been converted to sugar and that the process continues longer than necessary. It is therefore critical to obtain data concerning the amount of sugar and alcohol in the wort fast. It is possible to get these data, using absorbance data in the mid infrared region (MIR) and analyzing these statistically using PCA and Artificial Neural Network (ANN) to determine the amount of sugars and alcohol in the wort during the saccharification and the fermentation procedure. These optical techniques provide huge advantages because they can be easily adapted to the industrial equipment, providing real-time responses with a high specificity and sensitivity. By applying these techniques, the procedure of saccharification and fermentation can be modified in each brewing step to increase the quality of the wort and eventually of the beer. This routine analysis during processing can also be used for other liquid samples.

A main feature of ANN is its ability to learn from examples, without having been specifically programmed in a certain way. In the case of spectroscopy, satisfying results can be achieved when ANN is used with supervised training algorithms. The external supervisor (researcher) provides information about the desired response for the input patterns, i.e. where there is an "a priori knowledge" of the problem. A neural network can be defined as applying non-linear vector spaces between input and output. This is done through layers of neurons and activation functions, where the input values are added according to weight and "bias" specific, producing a single output value [12-14]. A network "feedforward" is progressive or shows no recursion, if the input vector and a layer formed by the values precede the output layer, as shown in figure 3.

Formally, the activation function of the i-th neuron in the j-th layer is denoted by $F_{i,j}(\times)$; its output itself, j, can be calculated from the output of the previous layer itself, j-1, the weights $W_{i,k,j-1}$ (the index k indicates the neuron connected to the preceding layer) and bias $b_{i,j}$ according to the following formula

$$s_{i,j} = F_{i,j}\left(b_{i,j} + \sum_{k} w_{i,k,j-1}\, s_{k,j-1} \right) \tag{6}$$

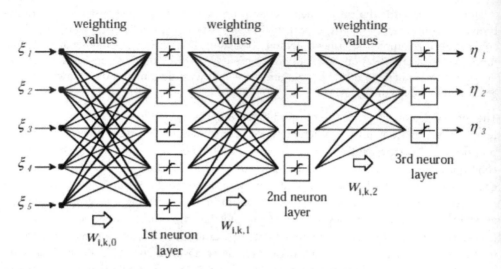

Figure 3. Schematic architecture of a neural network (perceptron multilayer).

The input and output values of the network being denoted by ξ_i and η_i respectively, the mapping can be determined due to a successive application of equation 6, which results for example in the following equation in the previous case:

$$\xi_i = F_{i,3}\left(b_{i,3} + \sum_{k=1}^{4} w_{i,k,2} F_{k,2}\left(b_{k,2} + \sum_{m=1}^{5} w_{k,m,1} F_{m,1}\left(b_{m,1} + \sum_{n=1}^{5} w_{m,n,0} \eta_n \right)\right)\right) \tag{7}$$

Since the choice of the activation function usually falls on the logistic sigmoid due to some of its mathematical properties (be class $C \infty$, for example), the above expression shows the relationship between ξ_i and η_i wich is defined by the weighing values and the bias. A very important characteristic of NN is its ability to learn, or the ability to reproduce the input-output pairs predetermined by properly adjusting the weights and the bias from training data and according to an adjustment rule. The method of a *"backpropagation"* rule is probably the best known training, and it is especially suited for progressive architectures. This rule is based on the successive application of the maximum slope algorithm determined from the first derivatives of the error between the desired outputs obtained by the parameters of the internal network. The backpropagation can be summarized in the following steps: (1) initialize the network parameters, $b_{i,j}$ and $w_{i,k,j}$ (2) select an entry ξ_i^p training data and form the pair (η_i^p, δ_i^p) , (3) calculate the error with a standard convenient Euclidean, e.g.

$$e = \sqrt{\Sigma_i(\delta_i^p - \eta_i^p)^2} \tag{8}$$

(4) Calculate the error derived from the above equation in relation to $b_{i,j}$ and $w_{i,k,j}$ (5) modify the parameters of the network according to the following rule and learning rate:

$$b_{i,j} \leftarrow b_{i,j} - \alpha \frac{\partial e}{\partial b_{i,j}} \quad and \quad w_{i,k,j} \leftarrow w_{i,k,j} - \alpha \frac{\partial e}{\partial w_{i,k,j}} \tag{9}$$

(6) Iterate steps (2) through (5) until a number of training cycles or stopping criteria has been achieved.[12, 13, 15, 16]

We can show in our case of beer analysis to which extent this processing technique is powerful. It has been applied widely in the interpretation of spectral data. In this case, an infrared absorption spectrum was obtained by Fourier Transform Infrared (FTIR) spectrometer [1, 2], the spectra is show in figure 4 and figure 5. In this case, the research objective was to provide a new method to determine the concentration of sugars and ethanol in beer wort during beer saccharification and fermentation in a short processing time. In our example, compounds of interest to be quantified can be separated into four main types of sugars present in the sample: glucose, maltose, maltotriose, dextrin (sugar chain length) and ethanol. It is important to note that the maltose binding is composed of two molecules of glucose, maltotriose three molecules of glucose sugar and that dextrins are composed of a large number of glucoses. Thus, the fundamental basis of these sugars is the same, the glucose, being differentiated only by the number of basic elements connected.

The absorption bands of these elements are expected to be so close that there is an overlap in the spectra, making the detection and quantification very complex. Figure 4 shows an example of absorption spectrum of a sample of ethanol, maltose 10%, and beer wort which contains some types of sugars. It is quite difficult to distinguish between the absorption spectra of the beer wort and the maltose, which contains certain types of sugars.

If we consider also the presence of ethanol (which has an absorption band in the same spectral region as the sugar) in the fermentation step, the procedure becomes even more complex. In figure 5 the extent of absorption during the fermentation step is shown, where the sample had initially all sugars without ethanol and ends up having only a part of dextrin (no fermentable sugar) and ethanol.

In this case, we first use the technique of principal component analysis in order to achieve a reduction of the number of variables to be analyzed. These spectra, which originally had about 1000 variables (wavelength where the absorbance is measured), can by these means represented by a few (two, in this case) variables, or principal components, with a high representation of information: 97.9%. The relationship between the two higher principal components is presented in figure 6 below.

Each spectrum of figure 5 is represented in figure 6 by a single point. In this new base of analysis, the wavelengths do not have any more significance, but the variance is important now. Each pair (PC1, PC2) represents a specific concentration of sugar and ethanol, which changes during the fermentation process. It is computationally feasible at this time, to apply an artificial neural network based on the values of the pairs for each of these points. For the first time, this experiment should be performed as previously described: as a case of a supervised NN e.g. a multilayer perception network). Therefore a method is required as the gold standard to calibrate or to train our neural network. One of the most widely accepted methods is the technique of HPLC (High Performance Liquid Chromatography). Using this technique, we can accurately quantify all the types of sugars of interest and the ethanol. The compounds of interest were measured using the standard method and assembling the ANN. In the neural network input (η_i) using ordered pairs of principal components and the output (ξ_i), the results obtained using the HPLC techniques show the amounts of compounds of interest, in this case, the sugars and the ethanol.

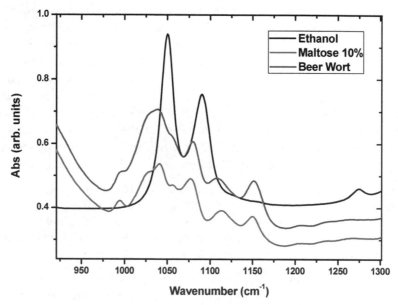

Figure 4. Absorption spectrum of a sample of ethanol, maltose 10%, and beer wort which contains certain types of sugars.

A certain part of the data (approximately 1/3), must be separated first in order to perform a further validation step. With 2/3 of the remaining data, the neural network is performed in the training stage following the equations and the structures described before, where the weight of each neural layer is adjusted in order to converge the network. The adjustment can be done as often as necessary, until the output (ξ_i) is as close to the true (real) value as required.

The training is complete, oncet the weight of the neurons has been adjusted, and the network has been converged with the desired error. The weight values should then be saved and stored before proceeding to the next step, which is the validation step: using the neural network to provide results of new spectra. With the data that were originally separated (1/3 data) and the values of the weights defined by the training stage, the neural network is run again. At this stage, note that the backpropagation system should not be executed. Simply use the matrix of the weights saved, and insert the data that have been separated for this validation step as inputs for the new network. Thus, the network will be performed only in the forward direction, supplying in a very short time of processing the output values ξ_i.

These output values are compared with the expected values using the HPLC technique, using a correlation curve between the two techniques. If these results are satisfactory, the process of mounting the system to quantify the compounds of interest is complete, and can be passed on for practical use.

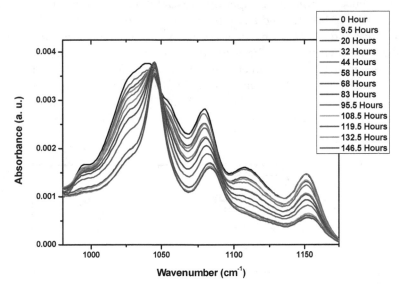

Figure 5. Absorption during the fermentation process.

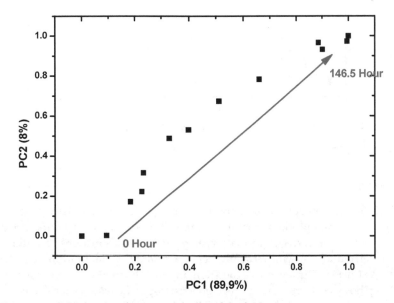

Figure 6. PC1 versus PC2 showing the time evolution of the fermentation process.

To be able to use spectroscopy with the neural processing requires using a standard method. We can simply use the PCA to reduce processing variables, entering the values of ordered pairs into the network, together with the weight values and collecting predetermined output results, in this case the amounts of sugars and ethanol. In the case of fermentation of the wort, using a number of principal components around three , a neural network comprising an input layer with 23 neurons and an output layer of 5 neurons , it is possible to quantify each type of sugar and ethanol with a quoted error of ± 0.2%. Here we exemplify our results showing the correlation between the value determined by the concentration of maltose using spectroscopy and HPLC technique (figure 7), where R^2 and the coefficient slope is 0.991 and 0.999 respectively. The results of a linear fit show a good agreement between the proposed new method and the standard procedure. This result allows the use of our technique in brewery, as it enables monitoring quality and making process control less time consuming.

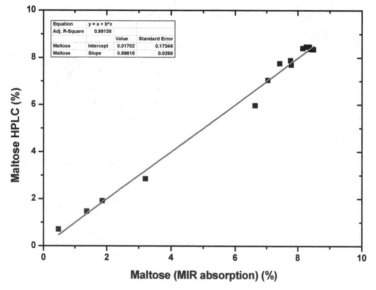

Figure 7. Correlation between the standard method (HPLC) and proposed procedure (MIR absorption).

4. Conclusion

In the analysis of spectroscopic data, not only the technique to obtain the values of different properties is important, but the correct mathematical processing of the data is actually the main issue to obtain the correct information. Especially the distinctions of multiple values which are correlated to a specific class of phenomena are the hide information that can be conveniently extracted. During our exposition in this chapter, we have concentrated in demonstrating how powerful the correct spectroscopy analysis can be when the first obtained data have been correctly arranged, allowing a mathematical procedure that treats the information as a whole instead of concentrations in individual values. Many techniques are today available for such procedures, but especially the Principal Component Analysis is quite powerful to be applied when spectral information is not restricted to a single wavelength, but rather to a large portion of the spectra.

We have concentrated on a relevant case where the UV-VIS portion of fluorescence spectrum is obtained and applied to determine its correlation with the postmortem interval in an animal model. The fluorescence in this case is subject to many effects due to the biological tissue modification as a natural evolution once the living metabolic action has been interrupted. This is clearly the case where biochemical modification causes alteration of spectrum as a whole and the attempt to concentrate the observation on individual features may fail. With the application of PCA to collect data, rich information patterns made a high correlation between extracted information and the real postmortem time interval possible. The classification of patterns and congregations of collections of information create a distinction into groups of distinct PMI. Even though we have used the method for PMI determination, the method has been shown to be as well powerful in applications in the field of cancer diagnostic, fermentation processing in beverage production, quality control in industry, identification of plagues and other features of interest in agriculture. The level of application of the PCA technique can go beyond the identification of pattern and correlation with values and can also provide specific quantification of individual chemical components of the system which is investigated.

To demonstrate this feature, we consider as an example the sugar quantification during beer production. These cases represent a bigger challenge to innumerous systems in several areas. Using the PCA procedure associated with a Neural Network (NN) we can quantify the composites in a sample, obtaining results comparably quickly. Here, we used the example of beer analysis. Using the MIR absorption spectroscopy of liquid samples, without any type of pre-procedure, we detected and quantified specific compounds (glucose, maltose, maltotriose, dextrin and ethanol) during the production of beer. The NN were used to determine the amount of these types of sugar and alcohol in the wort during the saccharification and fermentation. In the correlation between the values determined by the concentration of maltose spectroscopy with the HPLC technique we find the R2 and coefficient slope to be 0.991 and 0.999 respectively. Finally, the presentation of this chapter is to show the real power of the conjugation of spectroscopy techniques with data analyses. The field is clearly growing in diversity and importance.

Author details

E.S.Estracanholli, G.Nicolodelli, S.Pratavieira, C.Kurachi and V.S. Bagnato
Institute of Physics of São Carlos, University of São Paulo, SP, Brazil

Acknowledgement

The authors acknowledge the support provided by CNPq (INOF – INCT Program and scholarship) and FAPESP (CEPOF – CEPID Program and scholarship).

5. References

[1]   Socrates G. Infrared and raman characteristic: group frequencies: J. Wiley 2001.

[2]   Stuart BH. Infrared spectroscopy: Fundamentals and applications: J. Wiley 2004.

[3]   Bauman RP. Absorption spectroscopy: John Wiley 1963.

[4]   Lakowicz J. Principles of Fluorescence Spectroscopy. Springer 2006.

[5]   Ramanujam N. Fluorescence spectroscopy of neoplastic and non-neoplastic tissues. Neoplasia. 2000 Jan-Apr;2(1-2):89-117.

[6]   Estracanholli ES, Kurachi C, Bagnato VS. Tissue Fluorescence Spectroscopy in Death Time Estimation. *Forensic Pathology Reviews*: Springer Science 2011.

[7]   Pratavieira S, Andrade CT, Salvio AG, Bagnato VS, Kurachi C. Optical Imaging as Auxiliary Tool in Skin Cancer Diagnosis. *Skin Cancers - Risk Factors, Prevention and Therapy* 2011.

[8]   Estracanholli ES, Kurachi C, Vicente JR, de Menezes PFC, Silva OCE, Bagnato VS. Determination of post-mortem interval using in situ tissue optical fluorescence. Optics Express. 2009 May 11;17(10):8185-92.

[9]   Schieberle HDBWGP. Food chemistry: Springer 2009.

[10] Woodcock T, Downey G, O'Donnell CP. Better quality food and beverages: the role of near infrared spectroscopy. Journal of near Infrared Spectroscopy. 2008;16(1):1-29.

[11] Bamforth CW. Food, fermentation and Micro-organisms. : Blackwell Science 2005.

[12] Kohonen T. Analysis of a Simple Self-Organizing Process. Biological Cybernetics. 1982;44(2):135-40.

[13] Kohonen T. Self-Organized Formation of Topologically Correct Feature Maps. Biological Cybernetics. 1982;43(1):59-69.

[14] Kandel ER, Schwartz JH, Jessel TM. Principles of neural science: McGraw-Hill 2000.

[15] Lefebvre JCPNREWC. Neural and Adaptive Systems: Fundamentals through Simulations: John Wiley & Sons 1999.

[16] Marks RDRRJ. Neural smithing: supervised learning in feedforward artificial neural networks. : MIT Press 1999.

Permissions

The contributors of this book come from diverse backgrounds, making this book a truly international effort. This book will bring forth new frontiers with its revolutionizing research information and detailed analysis of the nascent developments around the world.

We would like to thank Muhammad Akhyar Farrukh, for lending his expertise to make the book truly unique. He has played a crucial role in the development of this book. Without his invaluable contribution this book wouldn't have been possible. He has made vital efforts to compile up to date information on the varied aspects of this subject to make this book a valuable addition to the collection of many professionals and students.

This book was conceptualized with the vision of imparting up-to-date information and advanced data in this field. To ensure the same, a matchless editorial board was set up. Every individual on the board went through rigorous rounds of assessment to prove their worth. After which they invested a large part of their time researching and compiling the most relevant data for our readers. Conferences and sessions were held from time to time between the editorial board and the contributing authors to present the data in the most comprehensible form. The editorial team has worked tirelessly to provide valuable and valid information to help people across the globe.

Every chapter published in this book has been scrutinized by our experts. Their significance has been extensively debated. The topics covered herein carry significant findings which will fuel the growth of the discipline. They may even be implemented as practical applications or may be referred to as a beginning point for another development. Chapters in this book were first published by InTech; hereby published with permission under the Creative Commons Attribution License or equivalent.

The editorial board has been involved in producing this book since its inception. They have spent rigorous hours researching and exploring the diverse topics which have resulted in the successful publishing of this book. They have passed on their knowledge of decades through this book. To expedite this challenging task, the publisher supported the team at every step. A small team of assistant editors was also appointed to further simplify the editing procedure and attain best results for the readers.

Our editorial team has been hand-picked from every corner of the world. Their multi-ethnicity adds dynamic inputs to the discussions which result in innovative outcomes. These outcomes are then further discussed with the researchers and contributors who give their valuable feedback and opinion regarding the same. The feedback is then collaborated with the researches and they are edited in a comprehensive manner to aid the understanding of the subject.

Apart from the editorial board, the designing team has also invested a significant amount of their time in understanding the subject and creating the most relevant covers. They scrutinized every image to scout for the most suitable representation of the subject and create an appropriate cover for the book.

The publishing team has been involved in this book since its early stages. They were actively engaged in every process, be it collecting the data, connecting with the contributors or procuring relevant information. The team has been an ardent support to the editorial, designing and production team. Their endless efforts to recruit the best for this project, has resulted in the accomplishment of this book. They are a veteran in the field of academics and their pool of knowledge is as vast as their experience in printing. Their expertise and guidance has proved useful at every step. Their uncompromising quality standards have made this book an exceptional effort. Their encouragement from time to time has been an inspiration for everyone.

The publisher and the editorial board hope that this book will prove to be a valuable piece of knowledge for researchers, students, practitioners and scholars across the globe.

List of Contributors

Wenjun Liang, Jian Li and Hong He
College of Environmental and Energy Engineering, Beijing University of Technology, Beijing, China

Kavian Cooke
Department of Mechanical Engineering, School of Engineering, University of Technology, Jamaica

Tatsuo Fujii, Ikko Matsusue and Jun Takada
Department of Applied Chemistry, Okayama University, Japan

Sam-Rok Keum, So-Young Ma and Se-Jung Roh
Korea University at Sejong Campus, South Korea

Leonardo M. Moreira
Departamento de Zootecnia (DEZOO), Universidade Federal de São João Del Rei (UFSJ), São João Del Rei, MG, Brazil

Juliana P. Lyon
Departamento de Ciências Naturais (DCNAT), Universidade Federal de São João Del Rei (UFSJ), São João Del Rei, MG, Brazil

Ana Paula Romani
Departamento de Química - Instituto de Ciências Exatas e Biológicas, Universidade Federal de Ouro Preto, Campus Morro do Cruzeiro, Ouro Preto, MG, Brazil

Divinomar Severino
Instituto de Química, Universidade de São Paulo, São Paulo, SP, Brazil

Maira Regina Rodrigues
Universidade Federal Fluminense, Polo Universitário de Rio das Ostras, Rio das Ostras, RJ, Brazil

Hueder P. M. de Oliveira
Centro de Ciências Químicas, Farmacêuticas e de Alimentos, Universidade Federal de Pelotas, Pelotas, RS, Brazil

Luxi Li and Xianbo Shi
Brookhaven National Laboratory, Upton, NY, USA

Cherice M. Evans
Department of Chemistry, Queens College – CUNY and the Graduate Center – CUNY, New York, NY, USA

Gary L. Findley
Chemistry Department, University of Louisiana at Monroe, Monroe, LA, USA

Denys Kurbatov and Anatoliy Opanasyuk
Sumy State University, Sumy, Ukraine

Halyna Khlyap
TU Kaiserslautern, Kaiserslautern, Germany

E.S.Estracanholli, G.Nicolodelli, S.Pratavieira, C.Kurachi and V.S. Bagnato
Institute of Physics of São Carlos, University of São Paulo, SP, Brazil

9 781632 384263